Direct Effect of European Law and the Regulation of Dangerous Substances

Environmental Technology

A book series edited by Richard F. Griffiths

Environmental Technology Centre, UMIST, UK

Volume 1

Water Sampling for Pollution Regulation

Keith D. Harsham

Volume 2

Direct Effect of European Law and the Regulation of Dangerous Substances

Christopher J. M. Smith

This book is part of a series. The publisher will accept continuation orders which may be cancelled at any time and which provide for automatic billing and shipping of each title in the series upon publication. Please write for details.

Direct Effect of European Law and the Regulation of Dangerous Substances

Christopher J. M. Smith
Her Majesty's Inspectorate of Pollution, Lancaster, UK

Routledge
Taylor & Francis Group
LONDON AND NEW YORK

First published 1995 by Overseas Publishers Association

2 Park Square, Milton Park, Abingdon, Oxon OX14 4RN
711 Third Avenue, New York, NY 10017, USA

First issued in paperback 2016

Routledge is an imprint of the Taylor & Francis Group, an informa business

The views expressed in this book are those of the author and not necessarily of Her Majesty's Inspectorate of Pollution.

British Library Cataloguing in Publication Data

Smith, Christopher J. M.
Direct Effect of European Law and the
Regulation of Dangerours Substances. –
(Environmental Technology Series, ISSN
1023-9650; Vol. 2)
I. Title II. Series
341.762

ISBN 978-2-88449-041-2 (hbk)
ISBN 978-1-138-96772-4 (pbk)

CONTENTS

INTRODUCTION TO THE SERIES

The generation of material wealth through industrial production carries with it inevitable impacts on the environment. The challenge to society is to reconcile these factors so as to achieve a high level of protection for the environment as a whole whilst continuing to enjoy the benefits of industrial activities. These objectives necessitate the development of an integrated approach to pollution control and environmental protection and are at the heart of the principle of sustainable development. Achievement of these goals calls upon contributions from many disciplines and professions, including technology, the law and regulatory arrangements, and environmental management.

The series will include specialist technology-based topics in relevant industry sectors, involving clean process design as well as abatement technologies, along with titles dealing with management and regulatory aspects of environmental protection. These will address the subject matter in the context of the need to develop the integrated approach which is currently the focus of many European regulatory requirements.

The series will be of use to operators of industrial processes as well as regulatory bodies and those involved in environmental consultancy, and some titles will be appropriate for degree-level courses.

FOREWORD

The rather arid title of this volume cloaks an issue of profound practical importance for individuals and institutions alike who are involved in the implementation of European Union directives or are affected by their provisions. The Doctrine of Direct Effect is a legal concept that derives from the relationship between the requirements of EU directives and the arrangements made by member states for their implementation. The essence of the doctrine is that individuals may rely upon the provisions of directives even where the member state has failed to make arrangements to implement them in national legislation, or has implemented them incorrectly. Provided that the provisions in question are clear, precise, and unconditional, direct effect can be relied upon.

Where individuals suffer loss as a result of such failure, they are able to seek damages against the relevant body. Liabilities in this respect may fall on enforcing authorities and other organisations created by actions of the Member State, including for example entities created by privatisation.

The author explores the implications of the doctrine in the context of EU directives concerning the environment. In particular, attention is focused on implementation of the Directive on pollution caused by certain dangerous substances discharged into the aquatic environment of the Community (76/464/EEC), commonly referred to as the Dangerous Substances Directive.

After briefly describing the institutions and legal structures of the EU, he examines a wide range of cases from many sectors of regulation in order to establish the general scope of the doctrine. This is followed by a detailed analysis of the implementation of the Dangerous Substances Directive, revealing numerous potential liabilities in respect of which the Doctrine of Direct Effect could be invoked.

This book explores a dimension of EU law which should be understood by regulators in all enforcing authorities, and which will also be of great importance to those who are regulated.

Richard F. Griffiths
UMIST, UK

1. INTRODUCTION

There is now considerable precedent in European Law concerning the so called doctrine of "direct effect".

The basis of the doctrine lies in the fundamental aspects of the Treaty of Rome and concerns the authority in Member States of treaty provisions and regulations, directives, decisions, recommendations and opinions coming from such provisions.

One important aspect of the doctrine is that responsibility for the implementation of European Law extends beyond governments to the national regulatory bodies themselves. The doctrine goes further in that it is argued that, even if implementation of European Law in a Member State by its government is deficient, the onus for the full and proper implementation of such law may still extend to regulators. Such deficiency may include incorrect as well as incomplete implementation and it is suggested that regulators may even find themselves liable for the payment of compensation if they fail to carry out their role correctly.

This monograph:

(1) examines the institutions and systems that are in place in the European Community with particular regard to those that effect the development and implementation of Community law;

(2) examines the development of the doctrine of "direct effect" and considers the importance of this to the agencies in England and Wales that are tasked with the control of discharges of dangerous substances into the aquatic environment; and

(3) examines the implementation, in England and Wales, of directives relating to the control of discharges of dangerous substances to water.

A consideration of the doctrine of direct effect would be of no significance if all Member States were fully and accurately to implement requirements of European Law into national law. Unfortunately this is not the case. Given below is a summary of a report prepared by the Commission of the European Communities in September 1992 on the level of implementation of directives relating to the environment. From this information it is clear that there is a considerable shortfall in the notification of implementing measures by Member States.

The report concluded that "the average success rate of 85% is quite good although delays in Italy and Greece give cause for concern."[1]

Apart from the unfortunate fact that no Member State had at that time implemented all the directives as required, this figure in no way takes account of any incomplete or faulty implementation. It assumes that notification of action, by the Member State to the Commission, constitutes full and accurate translation of directive obligations into national law.

A report prepared for Directorate General XI by Environmental Resources Management Limited in December 1993 concludes "Our research has shown a very wide range in the rate and level of implementation of the directives." [76/464 and its daughters]. "This reflects widely differing background conditions in Member

Implementation of environmental directives in the European Community.

	Directives Applicable on 31 December 1991	Directives for which measures have been notified	%
Belgium	92	75	81
Denmark	92	91	98
Germany	92	84	92
Greece	91	70	76
Spain	91	85	93
France	92	82	89
Ireland	91	77	84
Italy	92	55	59
Luxembourg	92	80	86
The Netherlands	92	88	95
Portugal	91	86	94
United Kingdom	91	77	85

States in terms of administrative legal systems and resources. The successful implementation of the requirements of Directive 76/464 and its daughter directives in all Member States still appears to be some way off and this fact needs to be explicitly addressed in any further policy developments in this area."[2]

It can be seen, therefore, that the issue of direct effect is one of real significance for all organisations with responsibility for regulating in any area covered by European Law.

As of 1 January 1994 the European Community ceased to exist, being replaced by the European Union. Most of the reference material used in this monograph pre-dates the formation of the European Union and as such the term "European Community" has been retained. References to the European Community should, where appropriate, be considered as references to the European Union.

2. METHODOLOGY

Much of the material on European case law reviewed in this monograph has been taken from European Court Reports (ECR) or Common Market Law Reports (CMLR). It is useful to consider the relevance of these sources in the reporting and recording of case law which subsequently provides us with precedent.

1. EUROPEAN COURT REPORTS

ECRs consist of:

(1) head notes which do not form part of the report;
(2) a review of the case by a judge rapporteur who is appointed by the court;
(3) the opinion of the Advocate General; and
(4) the decision of the Court.

The facts of the case are contained in the report by the judge rapporteur, but this is often not available for up to eighteen months after the case, or until such time as it has been translated into all the languages of the Member States.

The opinion of the Advocate General is based on a detailed review of the applicable case law. The decision of the Court will usually follow the opinion of the Advocate General, although in one case this did not happen and the Advocate General's comments were contrary and regarded as a dissenting opinion[3]. Under these circumstances it is not clear what reliance should be placed on either the Court decision or the reasoned opinion of the Advocate General.

The decision of the Court is in two parts. The first part comprises the reasoning or motivation for the decision. The second part consists of the Court Order. The relevance of the first part in relation to the second is generally regarded as unclear when looking at these decisions for the purpose of determining precedent.[4]

2. COMMON MARKET LAW REPORTS

These are a good source of otherwise unreported, and therefore unavailable, cases but they are not to be regarded as official reports. It is found that on occasions there are significant differences between CMLRs and ECRs[5]. As such, and because of the unofficial nature of CMLRs, one may not be able to rely on their content for determining precedent.

3. THE ROLE OF CENTRAL GOVERNMENT

Discussion in this monograph on the role of government departments in the implementation of European law is based primarily on interviews that the author has undertaken with members and ex-members of the Department of the Environment.

4. THE ROLE OF THE REGULATORS

Interview techniques were also used to gain information on how the National Rivers Authority (NRA) and the privatised water companies see their role in regulating discharges of dangerous substances into the aquatic environment.

5. THE VIEW OF INDUSTRY

To explore industry's perception of how the Dangerous Substances Directive[6] has been implemented in England and Wales a standard letter was sent to the following trade associations:

Chemical Industries Association
International Wool Secretariat
Textile Finishers Association
Confederation of British Industry
Metal Finishing Association

3. THE INSTITUTIONS OF THE EUROPEAN COMMUNITY

There are four main institutions of the European Community.

1. THE COMMISSION

The first institution is the Commission and this comprises the executive of the Community. It consists of independent appointees from the Member States, two from each of the five larger States, and one each from the seven others. It has responsibility for implementing the policy of the Community and to do this will draft proposals for legislation for the approval of the Council of Ministers. It also has responsibility for drawing up environmental action plans. By using information from Member States, on for example proposed national legislation, it will form a view as to whether community wide control in such areas is necessary.

Internally the Commission is divided into a number of separate directorates and it is Directorate General XI which has responsibility for environmental matters.

It is important when looking at the action of the Commission in the area of environmental protection to appreciate its original ambivalence toward environmental matters. One should remember that the original aim of the Commission was to remove barriers to free trade, with environmental protection measures being taken to avoid distortion within the internal market of the Economic Community. Such distortion could easily arise from disparate levels of environmental protection within the Member States. This ambivalence has to some extent been lessened by changes in the scope of the Community policies which now require the incorporation of environmental protection into its other areas of activity.

2. THE COUNCIL OF MINISTERS

The second institution of the Community is the Council of Ministers, which is a political body consisting of one appointed minister from each Member State. The appointee will vary depending on the issue under consideration.

There are many Councils; two examples are the Council for Trade and Industry, and the Council for Arts. The Council which considers environmental matters meets twice each presidency and is made up of environmental ministers from

---

Each presidency runs for 6 months, from January to June and July to December, passing from one country to the next in alphabetical order based on the country's name in its own language. New Member States will join at the end of the "presidency running order" and once each Member State has presided over the Council the order starts again at the letter "A". Problems arise when there are even numbers of Member States. This means that each Member State would preside each time during the same half of the year. Presiding during August to December is regarded as disadvantageous because this working period is shorter, due to national holidays. In order to avoid always disadvantaging the same countries, in this current round, the two countries that have presidencies in the same year will preside in reverse alphabetical order. In 1995, therefore, France will preside before Spain (*E*spania).

each of the Member States. These ministers are usually supported by the permanent representative from the foreign embassy and at least one senior, technically qualified, civil servant.

Although the Council has as its objective the furtherance of Community policy, it is clearly influenced by national political bias and as such the Council's voting procedure is crucial. As will be seen this voting procedure depends on which Treaty Article has been used for adoption of proposed legislation, there being two distinct systems.

The first is by unanimity of all the members of the Council.

The second is by qualified majority and involves a system of weighting whereby two of the larger States or five of the smaller States can be outvoted. This system allows environmental policy to be pushed forward more quickly and avoids the possibility of national veto by individual dissenting Member States. For example, the Directive on the control of large combustion plants was delayed for many years by the opposition of just two Member States[7].

As a result of proposed changes to the current system of vote weighting, which would follow future expansion of the European Union, it will be even more difficult for Member States to veto proposed European legislation. However, it is important to include some democratic process of debate in any system that allows for the wishes of any individual Member State to be overruled. Such democratic input is provided by the next Community Institution, the European Parliament.

3. THE EUROPEAN PARLIAMENT

The European Parliament is made up of 567 MEPs who are directly elected every five years by the electorates of the twelve Member States. The Parliament acts primarily as a consultative body, although it also has some supervisory powers over the Commission with regard to the Community budget. It does, however, ensure that decision making in the Community retains a certain amount of openness, a necessary function in view of the fact that all the dialogue between the Commission and the Council of Ministers is carried out in secret. The "democratic" nature of the involvement of the Parliament has been particularly relevant since 1979 when simple appointments gave way to direct election of representatives.

The 1986 Single European Act also significantly changed the Parliament's powers in relation to Community legislation. Under the "Cooperation Procedure", as set out in Article 149 of the Treaty (as amended by the Single European Act), the Parliament now has powers to propose amendments to, or to reject, any proposed legislation that is the subject of qualified majority voting. If proposals are rejected by the Parliament the measures could then only be adopted by the Council by unanimity. Any amendments proposed by the Parliament must be considered by the Commission and, if as a result these are inserted in the proposed legislation, the Council must accept these by qualified majority, or adopt some other proposals by unanimity, or let the proposals lapse. Ball and Bell[8] give an example of such a situation in relation to Directive 89/458 on Emissions from Small Cars, where the Parliament inserted a stricter test than was originally proposed by the Commission, or which was wanted by the majority of the Council. As a result of the Parliamentary consultation stage the Commission inserted the stricter test into its proposals. The Council as a result had a choice of agreeing to the proposals or not having any directive at all.

Because of the difference in the voting procedures, it is important whether a directive is adopted under Article 100a (qualified majority) or under Article 130s (unanimity). This question was addressed in the case of *Commission v Council*[9], which concerned the Directive on waste from the titanium dioxide industry. In this case the Council considered that the matter under consideration was primarily of environmental concern. It argued that the principal objective of the Directive was to reduce and eliminate pollution caused by the titanium dioxide industry. It proposed, therefore, that the Directive should be adopted under Article 130s (which does not allow for any input from the Parliament under the "Cooperation Procedure"). The Commission and Parliament felt that the matter was primarily concerned with the establishment of the internal market. The Commission presented studies which sought to demonstrate that differing pollution control standards, imposed on the titanium dioxide industry in different Member States, had a major impact on their costs. These had resulted in a price differential across the Community of 10–20% and it argued, therefore, that the Directive should be adopted under Article 100a. The European Court of Justice ruled that Article 100a should be used where the area of consideration related to the establishment of the internal market, and Article 130s should be used where the area of consideration related to environmental matters. It concluded that the goals of environmental protection and the removal of market distortion under this Directive were indivisible. However, in this particular case the Commission's view that the Directive should be adopted under Article 100a was upheld. No doubt the explicit reference to the harmonisation of pollution reduction programmes for the industry, both in the title and first Article of the Directive, weighed heavily with the European Court.

Despite the Council's stance in this case most European Community governments support the extension of qualified majority voting on all environmental matters. Some future extension of the use of the "Cooperation Procedure" should be expected.

4. THE EUROPEAN COURT OF JUSTICE

The next institution to be considered is the European Court of Justice. This consists of thirteen judges appointed by the Member States with reasoned opinions being submitted to the European Court of Justice by one of the six Advocates-General.

The European Court of Justice has ultimate supremacy in its interpretation of European law, including Treaty Articles and subordinate secondary legislation. To facilitate this function, Treaty Article 177 allows for national courts to refer any matter of European Community law to the European Court of Justice for its interpretation of the law. Ball and Bell[10] describe the approach by the European Court of Justice in interpreting European law as "exceptionally activist and creative" and as "purposive" in its approach to interpretation, looking more at the spirit rather than the letter of the law. The development of the doctrine of direct effect and supremacy of European law can be directly attributed to the creative attitude of the European Court of Justice.

5. THE COURT OF FIRST INSTANCE

Article 168a of the Maastricht amendment established the Court of First Instance to hear simpler cases with an appeals procedure to the main court. It may deal with appeals on points of law but does not have jurisdiction to deal with Article 177 references or actions against or between Member States or Community Institutions.

6. THE ECONOMIC AND SOCIAL COMMITTEE (ECOSOC)

"The Economic and Social Committee assists the Council and the Commission. It consists of 189 members, representing various sectors of economic and social life. It must be consulted before decisions are taken on a large number of subjects, and is also free to submit opinions on its own initiative."[11] Article 198 of the Maastricht Treaty amendment requires that "The Committee must be consulted by the Council and Commission where the Treaty so provides. Article 193 of the Treaty describes the ECOSOC as having an "advisory status" and consisting of "representatives of a cross-section of occupations and of the public".

In practice the Commission will submit its official proposals for, for example, a directive, to the ECOSOC which then sets up a working group of experts from various interest groups (eg from industry and environmental groups) to formulate an opinion or propose amendments to the draft directive. As stated above the ECOSOC is advisory only and any comments it makes on any draft European legislation do not have to be taken into account by the Commission.

7. THE EUROPEAN ENVIRONMENT AGENCY

Implementation of Regulation 1210/90, which provided for the setting up of a European Environment Agency, was delayed until agreement could be reached concerning the base of the agency. The three-year stalemate over the location of the European Environment Agency was broken on 29 October 1993 when the Council decided that the Agency's headquarters would be in Copenhagen[12].

The role of the Agency will initially be limited to the gathering and dissemination of information. There is commitment to review its remit after two years of operation and it is possible that its scope will develop into that of an Environmental Inspectorate with more of a regulatory role.

4. REVIEW OF THE IMPACT OF EUROPEAN LAW

1. COMMUNITY LAW AND INTERNATIONAL LAW

The concept of an international institution which is not a State but has legal personality in international and national law is not new.

It is commonplace for States to create institutions with international personality. Collins[13] refers to a "landmark" decision surrounding the advisory opinion of the International Court of Justice in the *Reparation for Injuries* case[14], on the relationship between Community law and international law. In the opinion of the International Court of Justice, Member States had created an international body, the United Nations, which had a legal personality distinct from its members, and which could pursue claims for damages done to itself and its employees. It was not a State nor did it have legal rights and duties the same as those of a State.

In the United Kingdom, the House of Lords clearly indicated, in their decisions surrounding the cases of *The International Tin Council*[15], that they recognised the concept of an international organisation which:

(1) had its own legal personality;
(2) had a legal personality independent of that of its members; and
(3) was not just an association of Member States without separate legal existence.

Collins goes on to argue that "the States which create an international institution must show an intention either expressed or implied to confer international legal personality on it." Certainly, Article 210 of the Treaty provides that the Community shall have legal personality, but this could simply refer to its legal personality within national legal systems of the Member States. However, if one considers the express treaty-making powers given to the Community by virtue of Articles 113, 114 and 238, there can be little doubt that the founder members of the European Community created an international organisation with a legally discrete personality.

Activities over which the European Community has powers are set out in the Treaties which establish the Community. There are three linked communities, the European Economic Community, the European Coal and Steel Community, and the Atomic Energy Community. It is the European Economic Community, established by the Treaty of Rome and amended by the Single European Act 1986, which is central to the control of, among other things, environmental matters laid out in Community policy. All references in this monograph to the "Treaty" are references to the Treaty of Rome.

Until the European Environment Agency is established in its role the Community has no effective "police force" on which to rely to oversee implementation of Community environmental law. As such, unlike national policy, Community policy needs to be made explicit in terms of its legislation and is therefore necessarily very detailed in its control. However, such Community law, particularly directives, will inevitably permit some discretion and it is necessary to look carefully at national implementing legislation to see whether Community policy goals are being achieved. For example, implementation of Community law may require the drafting of completely new pieces of national legislation, whereas on the other hand a Member State may choose to rely on extant legislation.

Collins states that "There can be no doubt that the United Kingdom Government has caused much unnecessary litigation and expenditure of much unnecessary cost by taking the position that implementation of the Equal Pay and Treatment Directive required only the continuation of, or minor amendment to, the Equal Pay Act 1970 and the Sex Discrimination Act 1975, rather than new laws based specifically on the directives."[16] Collins expresses a firm view that the United Kingdom took the wrong course of action in relying on small amendments to extant legislation. However, with the propensity in the United Kingdom toward "framework" legislation, fleshed out with control under Statutory Instruments, there is a great tendency toward the reliance on extant statutes. It is also true that much of the environmental legislation in the United Kingdom lends itself to adaptation to implement European obligations, and this is particularly true in the area of aquatic pollution prevention.

2. INSTRUMENTS OF COMMUNITY LAW

Community law can be conveniently considered as a hierarchy extending from the Treaty of Rome, as modified by the Single European Act 1986, to individual "instruments" arising from the treaty such as regulations and directives. These are considered below.

2.1. *The Treaty of Rome*

Community law is a system of law created by treaty. This essentially independent legal system is given effect by the creation of Community Institutions and by its recognition in national legal systems. Community law, created by treaties setting up such institutions, operates within and overlaps national law but is nevertheless superior in that it restricts the sovereignty of Member States.

It is the treaty setting up the European Economic Community that is of particular relevance to this discussion. Article 5 of the Treaty requires that:

"Member States shall take all appropriate measures, whether general or particular, to ensure fulfilment of the obligations arising out of this Treaty or resulting from action taken by the institutions of the Community.

They shall facilitate the achievement of the Community's tasks.

They shall abstain from any measure which could jeopardise the attainment of the objectives of this Treaty."

Since the United Kingdom's accession to the European Community in 1973, the Treaty of Rome has been an important source of European legislation on pollution control. However, although the Treaty is having increasing impact on all areas of activity in Member States, environmental matters have no specific mention in its Articles. Prior to 1987 the legal basis for environmental control came from two Articles in the Treaty of Rome, Articles 100 and 235.

Article 100 states:

"The Council shall, acting unanimously on a proposal from the Commission, issue directives for the approximation of such provisions laid down by law, regulations or administrative action in Member States as directly affect the establishment or functioning of the common market."

Article 235 states:

"If action by the Community should prove necessary to attain, in the course of the operation of the common market, one of the objectives of the Community and this Treaty has not provided necessary powers, the Council shall, acting on a proposal from the Commission and after consulting the European Parliament, take the appropriate measures."

The first of these Articles concerns the issuing of directives by the Commission for the purpose of establishing the common market. The second is essentially a "catch all" giving the Community the power to take action where no other Treaty provision is seen to be appropriate.

It was not until the Paris Summit of 1972 that the Heads of State agreed that economic expansion should go hand in hand with improvement in the quality of life, thereby giving birth to the Community Action Programmes on the environment.

Community action on environmental issues using Articles 100 and 235 was endorsed by the European Court of Justice in cases brought by the European Commission against Italy, concerning the directive on the biodegradability of detergents and the directive on maximum sulphur content of liquid fuels. The Court gave identical rulings stating that "As regards the observations of the Italian Government concerning the powers of the Community in the matter, it should be observed that the directive has been adopted not only within the Programme of Action of the Communities on the Environment; it also comes under the General Programme for the elimination of technical barriers to trade which result from disparities between the provisions laid down by law, regulation or administrative action in Member States, adopted by the Council on 28 May 1969. In this sense it is validly founded upon Article 100. Furthermore it is by no means ruled out that provisions on the environment may be based upon Article 100 of the Treaty. Provisions which are made necessary by considerations relating to the environment and health may be a burden upon undertakings to which they apply and if there is no harmonization of national provisions on the matter, competition may be appreciably distorted."[17]

2.2. *The Single European Act 1986*

As discussed above, before the Single European Act came into force in 1987 there were no specific provisions in the Treaty of Rome allowing the introduction of measures for the protection of the environment. Instead, and in order that environmental protection measures were not allowed to prevent the setting up of the internal common market, measures needed to harmonise pollution control were taken under Articles 100 or 235.

In 1987 the Single European Act amended the Treaty to provide specific powers to implement EC environmental policy. These powers are now contained in Articles 130r, 130s and 130t of the Treaty.

Article 130r states:

"1. Action by the Community relating to the environment shall have the following objectives:

(i) to preserve, protect and improve the quality of the environment;

(ii) to contribute toward protecting human health;

(iii) to ensure a prudent and rational utilization of natural resources.

2. Action by the Community relating to the environment shall be based on the principles that preventive action should be taken, that environmental damage should as a priority be rectified at source, and that the polluter should pay. Environmental protection requirements shall be a component of the Community's other policies.

3. In preparing its action relating to the environment, the Community shall take account of:

(i) available scientific and technical data;
(ii) environmental conditions in the various regions of the Community;
(iii) the potential benefits and costs of action or lack of action;
(iv) the economic and social development of the Community as a whole and the balanced development of its regions.

4. The Community shall take action relating to the environment to the extent to which the objectives referred to in paragraph 1 can be attained better at Community level than at the level of the individual Member State. Without prejudice to certain measures of a Community nature, the Member States shall finance and implement other measures.

5. Within their respective spheres of competence, the Community and the Member States shall cooperate with third countries and with relevant international organizations. The arrangements for Community cooperation may be the subject of agreement between the Community and the third parties concerned, which shall be negotiated and concluded in accordance with Article 228.

The previous paragraph shall be without prejudice to Member States' competence to negotiate in international bodies and to conclude international agreements."

Article 130s states:

"The Council, acting unanimously on a proposal from the Commission and after consulting the European Parliament and the Economic and Social Committee, shall decide what action is to be taken by the Community.

The Council shall, under the conditions laid down in the preceding subparagraph, define those matters on which decisions are to be taken by a qualified majority."

Article 130t states:

"The protective measures adopted in common pursuant to Article 130s shall not prevent any Member State from maintaining or introducing more stringent protective measures compatible with this Treaty."

To summarise, Article 130r sets out the European Community's policy on environmental protection. The objectives are the preservation, protection and improvement of the environment, using the principles of preventive action, that environmental damage should be rectified at source, and that the polluter should pay. Article 130r also requires the European Community to take environmental protection requirements into account in its other policies. It does, however, introduce the idea of subsidiarity, stating that the community shall take action relating to the environment to the extent to which the objectives can be attained better at Community level than at the level of the individual Member States. Member States are required to finance and implement other measures.

Article 189 of the Treaty provides the Community with the powers to issue legal instructions to its members. It states:

"In order to carry out their task the Council and the Commission shall in accordance with the provisions of the Treaty, make regulations, issue directives, take decisions, make recommendations or deliver opinions.

A regulation shall have general application. It shall be binding in its entirety and directly applicable in all Member States.

A directive shall be binding, as to the result to be achieved, upon each Member State to which it is addressed, but shall leave to the national authorities the choice of form and methods.

A decision shall be binding in its entirety upon those to whom it is addressed.

Recommendations and opinions shall have no binding force."

2.3. *Regulations*

Ball and Bell[18] describe Regulations as "legislative acts of general application" and "normally directly effective, as long as they are sufficiently precise."

An interesting development in the use of regulations is that of voluntary "codes", for example the voluntary regulation on eco-labelling[19]. The provisions of these regulations are directly effective in that individuals in any Member State wishing to subscribe to the schemes must comply with the terms of the regulation. However, participation is (at least for the time being) voluntary.

2.4. *Directives*

Unlike Community regulations, which are directly applicable, a directive is binding on the Member State to which it is addressed, with regard to the objectives to be achieved, but leaves the national authorities free to choose how it is to be implemented. As discussed below, a Member State cannot avoid this obligation by failing to adopt, by the appropriate deadline or in an appropriate manner, the necessary measures as required by the directive.

2.5. *Decisions*

Decisions are binding on the individual to whom they are addressed. They are rare in environmental law having been used mainly for gathering information, although precautionary steps concerning Chlorofluorocarbons were taken in the 1980's in the form of a decision[20].

2.6. *Recommendations and Opinions*

Recommendations and Opinions are not binding and are used as tools of persuasion.

3. THE MAASTRICHT AMENDMENT

An intergovernmental Conference to amend the Treaty was held in Maastricht, the Netherlands, in December 1991. Following this meeting, further changes to Treaty provisions dealing with the environment were agreed. These changes, together with changes to other parts of the Treaty, have now been ratified by Member States and are to be regarded as being in force.

Articles 2 and 3 of the Treaty list the objectives of the Community and specific action to be taken in the achieving of those objectives. Following the Maastricht amendments those objectives include for the first time reference to "Sustainable and non-inflationary growth respecting the environment." This, together with the European Community's other policy objectives, is to be achieved by "a policy in the sphere of the environment."[21]

The amendment also makes changes to Article 130r of the Treaty. This extends the list of environmental policy objectives to include the promotion of measures at international level to deal with regional or worldwide environmental problems. Additionally it requires that European Community environmental policies must in future be based on the precautionary principle. This means that the European

Community must not wait for incontrovertible evidence of enviromental damage before implementing environmental protection measures but must be proactive in its prevention of such harm.

Finally, under Article 130r, the wording governing the relationship between European Community environmental policy and its other policies, has been strengthened. The new wording provides that environmental protection must be integrated into the definition and implementation of other European Community policies.

The Maastricht amendments to Article 130s provide that, as a rule, proposed action on environmental protection shall be adopted by qualified majority. This is subject to the proviso that measures of a mainly fiscal nature, concerning town and country planning, land use measures and measures which relate to energy supply, will still be subject to adoption by unanimity. These measures could still be adopted by qualified majority if the Council of Ministers agree and these amendments should speed up the adoption of new environmental protection measures. However, the scope of exemptions is so wide that it is likely that a significant number of European Community environmental protection measures will still have to be adopted by unanimity.

It is generally assumed that majority voting results in tighter environmental standards than do decisions taken by unanimity. This is not always the case as was seen in the adoption of at least two directives. Tighter controls suggested by a minority of Member States in directives on the marketing and use of penta–chlorophenol[22] and on vehicle emissions[23] were both overruled by the dissenting majority[24].

Article 130t has also been amended by the Maastricht agreement, requiring that Member States notify the Commission where stricter national environmental measures have been adopted. This will enable the Commission to see whether such national legislation is consistent with the Treaty and to judge whether the introduction of more stringent European Community legislation is required.

5. IMPLEMENTATION OF EUROPEAN LAW IN ENGLAND AND WALES

1. THE FORMULATION OF DIRECTIVES

Proposals for directives may come from individual Member States or, as sometimes happens, following a meeting of ministers at a ministerial seminar. For example, in the case of the proposed Integrated Pollution Prevention and Control Directive, the impetus for action was provided by the enactment, in the United Kingdom, of the Environmental Protection Act 1990.

Such proposals are passed to the Commission who will be responsible for drafting the required implementing documents.

Stages of development of a directive would typically be as follows[25]:

1. a draft directive is prepared with an appropriate preamble covering all of the substantive points in the Treaty Articles, together with an explanatory memorandum;
2. a meeting is convened within Directorate General XI to discuss the proposals;
3. the draft is circulated to all other interested Directorates General;
4. copies of the draft are circulated to all Member States with an invitation to a meeting to discuss the proposals;
5. following this meeting an amended draft is prepared and re-circulated;
6. the amended draft is sent to relevant non-governmental organisations and industry groups for comment;
7. the amended draft, produced as a result of this round of consultation, is checked with the legal staff of Directorate General XI;
8. a competitiveness and employment impact statement is produced and formal approval of other Directorates General sought.

By this stage a document has been prepared which is ready to be formulated into a Commission proposal. This involves the following steps:

1. the draft is translated into all Community languages;
2. the draft is submitted to the Commission with a speaking note for the Commissioner explaining the benefits of the proposals, and the envisaged costs;
3. upon adoption by the Commission, the draft is published in the Official Journal of the Community, and copied to the European Parliament and the Council.

At this point it is appropriate to consider the progress of the proposals separately through the European Parliament and the Council.

The Parliament.

1. The rapporteur is contacted before the first meeting of the Committees that have been set up to review the proposal.

2. An attempt is made by a representative of the Commission to obtain a list of concerns which the rapporteur may have together with any comments made by industry, environmental groups etc. The Commission will then prepare responses to the identified concerns.
3. Meetings of the European Parliament Committees follow with responses being produced by the Commission to cover any issues brought up at these meetings.
4. A draft is then prepared which takes account of amendments proposed by the Committees and approved by a vote of the European Parliament. This constitutes the amended Commission proposal in draft form.

The Council.

1. The Council will announce plans to discuss the Commission's proposal at its Working Group on the Environment and a speaking note is prepared for this explaining the benefits that are anticipated from the proposal.
2. The Council Working Group (which is made up of officials from each of the Member States) will now discuss the proposals and consult with interested parties and a working group of national experts set up by the Member States. The Commission will be called upon to defend its proposals. However, if the Working Group Chairman considers that there is a majority of Member States wishing to modify any point, the Commission may be asked to suggest changes to its position.
3. The Working Group Chairman will now finalise the text resulting from the Group discussions and try to get agreement from the Member States for the revised text.
4. The Council text then goes before a meeting of the Committee of Permanent Representatives of the Member States (COREPER) to remove any outstanding points of contention. The group will finally decide what points remain for consideration by the Council of Environment Ministers.
5. The Council will hear statements by any Member State with outstanding concerns and a vote will be taken on the final text.

A meeting is then convened of legal linguists who will ensure that the meanings of the directive are the same in all Community languages. The directive is then published in the Official Journal of the European Communities.

2. CIRCULARS

The transposition of a directive into national law does not necessarily require the provisions of the directive to be enacted in precisely the same words in a specific legal provision; a general legal context may be sufficient if it ensures the full application of a directive in a sufficiently clear and precise manner. However, the European Court of Justice ruled in the case of *EC Commission v Netherlands*[26], which concerned the protection of wild birds, that there are administrative practices which, by their nature, can be changed at will by the authorities of the Member State and cannot, therefore, be regarded as constituting proper compliance. The Commission issued a Reasoned Opinion in July 1980 informing the

United Kingdom government that its practice of using circulars to implement the requirements of directives was not acceptable.

The status of Department of the Environment Circulars has always been unclear. They have no formal statutory basis, although in one case a planning inspector was held, by the planning authority, to have erred in law by failing to take account of Circular 1/85 on "The Use of Conditions in Planning Permissions"[27]. They are, however, regarded as a means of providing instructions to the relevant authorities. For example, Department of the Environment Circular 7/89 on "The implementation of European Community directives on pollution caused by certain dangerous substances discharged into the aquatic environment", begins "We are directed by the Secretaries of State for the Environment and for Wales to draw to the attention of the relevant authorities European Directive 76/464/EEC . . ."[28]

This rather informal style changes in the body of the above Circular which continues "All discharges liable to contain . . . [List I] substances must be authorised [sic] by the competent authority" and "The emission standard must determine for the substance concerned both the maximum concentration and the maximum quantity over a period of time, that are permissible in the discharge . . ."

Circulars are also used to promulgate the government's policy and interpretation of directives. For example Circular 7/89 continues "... competent authorities should seek to ensure as a minimum requirement that the appropriate Community quality objective set down in the directive concerned is met for all relevant discharges. Quality objectives should not, though, be allowed to become rigid and inflexible; the aim is to move progressively toward higher goals, including the minimisation of inputs of the most dangerous substances..."

Similarly, the Circular contains a working definition of the idea of "substantially increased handling capacity". (Several daughter directives require the use of "best technical means available" for abatement of releases of List I substances which arise from new or substantially increased discharges). The Circular explains, "The term substantially is not defined in the directives, but should be interpreted as an overall increase of 20% or more in the handling capacity of the particular substance."

Circulars are seen, therefore, as a flexible means of instruction, allowing statements of future policy, providing working interpretations, and setting out responsibilities and duties.

This semi-formal approach is not, however, accepted by the Commission as being an adequate means of implementation, as there is no clear means of enforcement of the contents. The very flexibility of circulars that makes them so attractive to Civil Servants makes them most unattractive to the European Commission[29].

3. ACTS OF PARLIAMENT

There are three main pieces of legislation that impact on the control of dangerous substances to the aquatic environment.

The first is the Water Industry Act 1991 which gives the water companies powers to (among other things), control discharges to sewers and hence their own sewage works. The role of the water companies with regard to the regulation of dangerous substances is rather ambivalent. Some of the prime concerns of the companies are to give a financial return to their shareholders, to encourage further investment, and thus enable expansion of the business. As such they are interested in attracting discharges of trade waste to sewer rather than discouraging discharges by the application of strict regulation.

However, they have themselves to comply with consent limits set on the outfall of their own sewage treatment works, and to comply with these they need to exercise some control over receipt of trade waste into the sewers. Such control is exercised using the provisions of the Water Industry Act 1991 although, as discussed below, List I substances are controlled by the Secretary of State.

The second piece of important legislation is the Water Resources Act 1991. This gives the National Rivers Authority (NRA) statutory powers to regulate discharges to controlled waters. Such discharges may be direct from industrial plant or from sewage works receiving trade wastes.

The third important Act of Parliament is the Environmental Protection Act 1990 and this removes many sources of List I substances from the control of the two previously discussed Acts. The Environmental Protection Act 1990 gives powers to the Chief Inspector of Her Majesty's Inspectorate of Pollution (HMIP) to regulate all releases from the potentially more environmentally polluting processes. The control of the release of List I substances to sewer and directly to controlled waters are just two facets of such regulation.

4. DIRECTIONS AND INSTRUMENTS

Much of the primary legislation in the United Kingdom, in the area of environmental protection, is in the form of a basic legislative outline. This legislation gives the Secretary of State for the Environment powers to issue other legally binding instructions thereby building on the basic framework.

The advantage of such a system is that Acts of Parliament can be kept (relatively) simple and brief. To incorporate all provisions in an Act would result in such a complex document that Parliament would be unable to engage in sensible debate on its contents. It is also much simpler to amend subordinate legislation, such as Directions or Statutory Instruments, which are issued using powers contained in the primary statute. It is also simpler to get such amendments through Parliament as much of this class of legislation passes through automatically unless specifically challenged or objected to by Members of Parliament.

There are several important pieces of subordinate legislation that impact in the area of regulation of List I substances to the aquatic environment.

First, the prescribed processes and regulations[30], which are issued under provisions of the Water Act 1989[31] and the later consolidating legislation, the Water Industry Act 1991[32]. These specify those substances which, when discharged to sewer, are regulated by the Secretary of State. HMIP exercises such control on behalf of the Secretary of State.

A further set of regulations[33] are those issued by the Secretary of State using powers contained within the Water Act 1989[34] and the later consolidating legislation, the Water Resources Act 1991[35]. These are, essentially, instructions to the private water companies and the National Rivers Authority concerning the provision of information and the attainment of water quality standards. More recently, regulations establishing a new system for classifying river quality have been issued[36] introducing statutory water quality objectives to replace non-statutory objectives introduced in the late 1970's.

The final set of regulations to be considered here are those issued using powers contained within the Environmental Protection Act 1990[37], and these set out those processes which require authorization from HMIP in order to operate legally[38].

Details of regulatory powers are described more fully below. It is by the use of such powers, contained within these Acts of Parliament and their subsequent secondary legislation, that discharges of List I substances in England and Wales are regulated. Unfortunately, this piecemeal development of legislation has led to some important discrepancies in the definition of those substances which are the subject of control, and these are summarised in the following table.

A comparison of the regulation of List I substances.

SUBSTANCE	DIRECTIVE DEFINITION	PRESCRIBED TO SEWER	PRESCRIBED UNDER EPA90[3]
Cadmium	Cadmium and its compounds	Cadmium and its compounds	1000 grammes of Cadmium and its compounds
Mercury	Mercury and its compounds	Mercury and its compounds	200 grammes of Mercury and its compounds
Pentachlorophenol	PCP and its salts	PCP (amended to PCP and its compounds)	350 grammes of PCP and its compounds
DDT	The sum of isomers DDT[1]	DDT	5 grammes of all isomers of DDT
HCH	The isomers of HCH	Gamma HCH	20 grammes of all isomers of HCH
Drins	Aldrin, Dieldrin, Endrin, Isodrin	Aldrin, Dieldrin, Endrin	2 grammes of Aldrin 2 grammes of Dieldrin 1 gramme of Endrin
Chloroform	Chloroform	Chloroform[2]	Not included
Perchloroethylene	Perchloroethylene (discharges > 30kg/ annum)	Perchloroethylene (discharges > 30kg/ annum)	Not included
Trichloroethylene	Trichloroethylene (discharges > 30kg/ annum)	Trichloroethylene (discharges > 30kg/ annum)	Not included
Carbon Tetrachroride	Carbon Tetrachloride	Carbon Tetrachloride	Not included
Hexachlorobutadiene	Hexachlorobutadiene	Hexachlorobutadiene	20 grammes of Hexachlorobutadiene
Hexachlorobenzene	Hexachlorobenzene	Hexachlorobenzene	5 grammes of Hexachlorobenzene
1,2 Dichloroethane	1,2 Dichloroethane (discharges > 30kg/ annum)	1,2 Dichloroethane	2000 grammes of 1,2 Dichloroethane
Trichlorobenzene	All isomers of Trichlorobenzene	Trichlorobenzene	75 grammes of all isomers of Trichlorobenzene

1 1,1,1-trichloro-2,2 bis (p-chlorophenyl) ethane
1,1,1-trichloro-2 (o-chlorophenyl) -2- (p-chlorophenyl) ethane;
1,1,1-dichloro-2,2 bis (p-chlorophenyl) ethylene
1,1,1-dichloro-2,2 bis (p-chlorophenyl) ethane

2 Arising from any process for the production of chlorinated organic chemicals or any process for the production of paper pulp

3 All masses given are maximum permitted quantities below which processes are exempt from control under the provisions of the Environmental Protection Act 1990 (provided releases to air and land are trivial).

The consequences of these anomalies are that not all List I substances are controlled by all pieces of legislation. It therefore depends on which control mechanism applies as to whether the requirements of the relevant directive are met. For example, a discharge to sewer of trichlorobenzene from a prescribed process (controlled under the provisions of the Environmental Protection Act 1990) would potentially be regulated in a different manner than the same discharge arising from a non-prescribed process (which would be controlled under the provisions of the Water Industry Act 1991). These different regulatory systems are explained more fully later.

The control philosophies of the different pieces of legislation are also different. To illustrate this, under the provisions of the dangerous substances daughter directives all new plants discharging List I substances are to be authorized only if they employ the "best technical means available", when that is necessary for the elimination of pollution. Where, for technical reasons, the standards set do not correspond to those achievable using the best technical means available, the Member State is to notify the Commission, before authorization is given, of those reasons. However, under the provisions of the Environmental Protection Act 1990, substances that are not prescribed (those substances that do not appear in Schedule 5 to Statutory Instrument 1991 No. 472) need only be rendered harmless. The requirement to eliminate (where possible) only applies to prescribed substances. This gives rise to a clear discrepancy in the level of regulation where List I substances are not prescribed for the purpose of control under the provisions of the Environmental Protection Act 1990.

5. THE VIEW OF THE NATIONAL COURTS

Enforcement of the above-mentioned pieces of primary and secondary legislation is carried out using the hierarchical court legal system which ultimately terminates at the House of Lords.

A summary of this hierarchy and the routes of appeal is shown in Figures 1 and 2.

The interpretation of all legislation is ultimately the responsibility of these Courts and as such their views on the implementation of European law are of the utmost importance. The attitude of National Courts toward the doctrine of direct effect is discussed below, but first we need to consider the available mechanism by which the Courts formulate their understanding of European law.

The most important measure used by the national courts is under Article 177 of the Treaty which allows questions of interpretation of European legislation to be referred to the European Court of Justice. This is particularly useful to the individual citizen when other means, such as appeals, have been exhausted. This is because the court of last instance is *required* in all cases to refer points which require clarification to the European Court of Justice.

The European Court of Justice operates an "open door approach" and has only on one occasion refused to give a ruling where it felt that proceedings had been artificially created[41].

When considering if an Article 177 reference is needed the national court must consider that a preliminary ruling is necessary to enable it to give a judgement. Guidelines were laid down on this by the European Court of Justice in case *283/81 Cilfit*[42]. In this case the Italian Court of Cassation asked the European Court of Justice whether the obligation to refer a question under Article 177 was con-

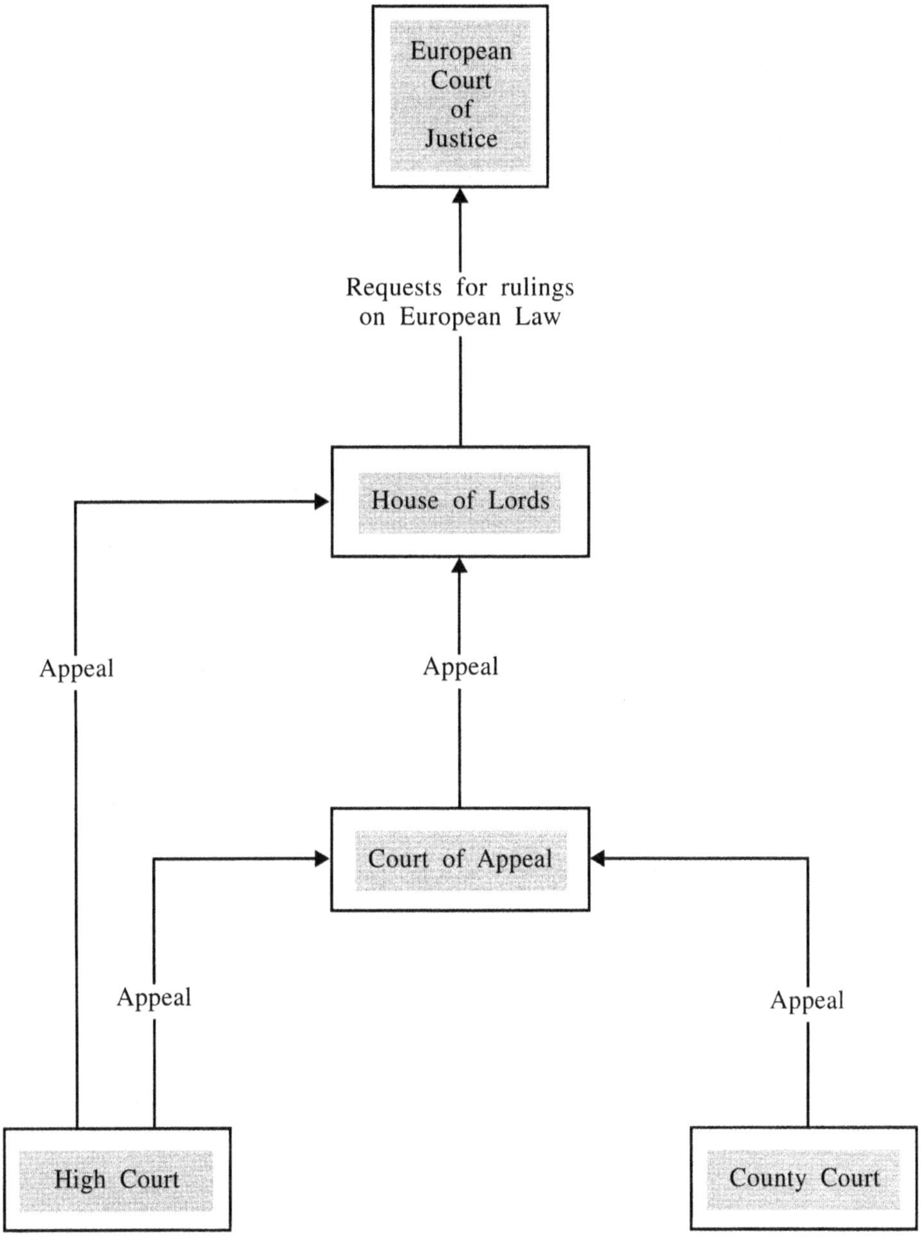

Figure 1 Principal Civil Courts[39]

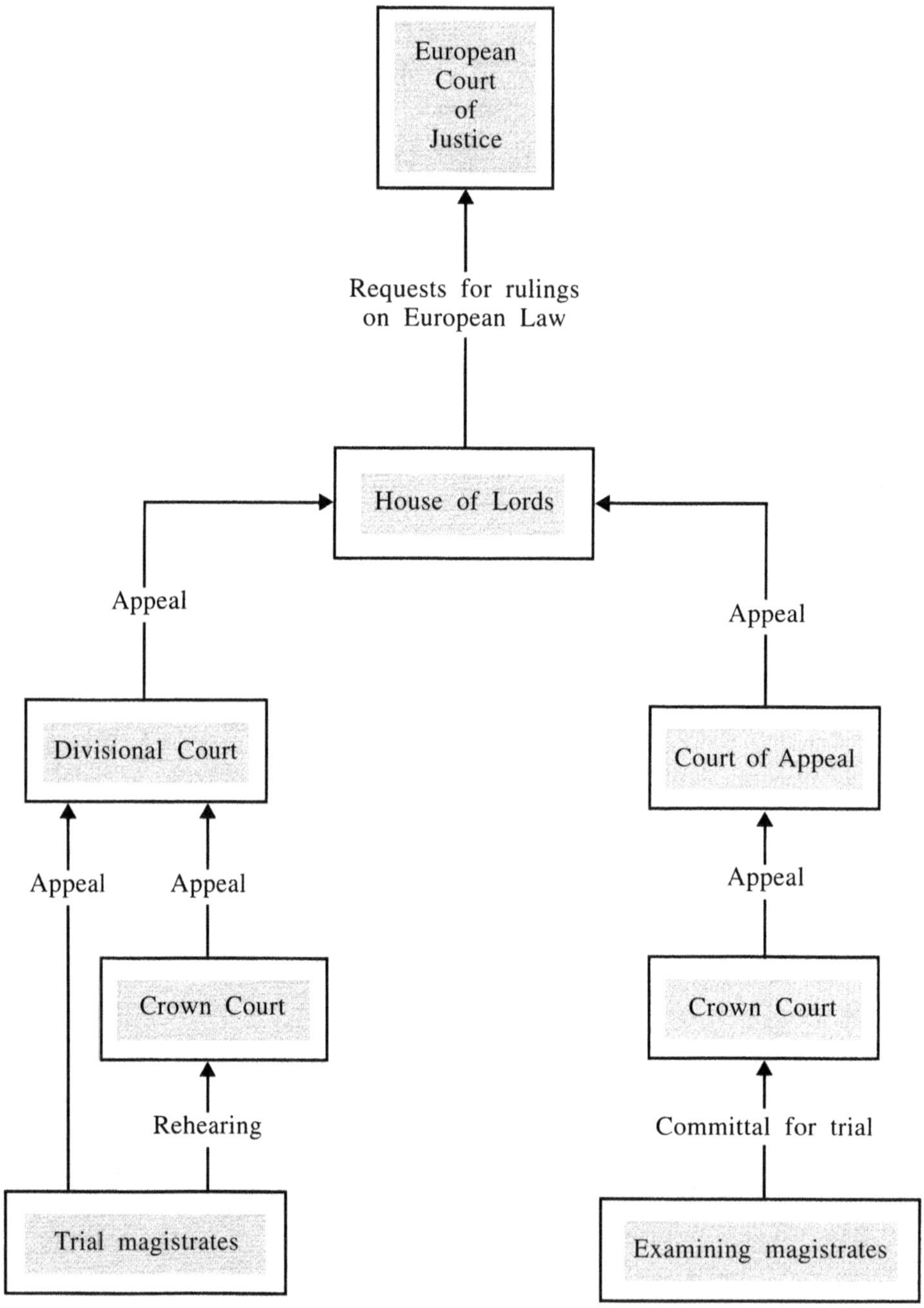

Figure 2 Principal Criminal Courts[40]

ditional on there being some reasonable doubt about the interpretation of the Community instrument involved. The European Court of Justice ruled that necessity for such a referral would not be established if:

1. the question of EEC law was irrelevant; or
2. the provision had already been interpreted by the European Court of Justice; or
3. the correct application was so obvious as to leave no room for doubt (*acte clair*).

If, after taking account of these limitations, the national court still considers that a referral is necessary, it will suspend its proceedings pending a ruling from the European Court of Justice. The European Court of Justice will only give a judgement in the context of the question referred and as such it is important for the national court to ask the right question.

Such a judgement is binding on the national court and defines the meaning of the rule of Community law which should have been applied at its coming into force. It acts therefore retrospectively, although interestingly in a few cases where major economic reasons exist (for example in compensation cases) the ruling has been declared not to be retroactive. However, even after a ruling, if a Member State insists on derogating from the line of authority established by the European Court of Justice, there is little in practice that can be done to prevent it from so doing[43]. This has been seen in the rulings of the French Conseil d'Etat in the case of *Minister of the Interior v Daniel Cohn-Bendit*[44] (described below).

Consideration as to whether a directive could be directly effective was central to the case of *Van Duyn v Home Office*[45]. In this case Miss Van Duyn had been refused entry into the United Kingdom because of her association with Scientology. One of the questions raised was whether a Council directive, coordinating special measures concerning the movement and residence of foreign nationals on the grounds of public policy, public security, or public health, was directly effective so as to confer rights on individuals. The Directive was intended to limit the discretionary powers of authorities in Member States controlling entry and expulsion of foreign nationals.

The United Kingdom government had argued its case before the European Court of Justice based on Article 189, which distinguishes between the effects to be ascribed to regulations and directives. The United Kingdom reasoned that it must be presumed that the Council, when issuing a directive rather than a regulation, must have as its intention that the Directive should have effects that differ from those of a regulation. The European Court of Justice ruled, however, that it would be incompatible with the binding effect attributed to a directive by Article 189, if the possibility that obligations placed on a Member State by such a directive could not be invoked by individuals in their national courts[46].

Nevertheless, the French Conseil d'Etat in the case of *Minister of the Interior v Daniel Cohn-Bendit* refused to follow the case of *Van Duyn v Home Office* and, in a much criticised judgement, held that directives could not be invoked by individuals. Consequently, Mr Daniel Cohn-Bendit was prevented from invoking the Free Movement of Persons Directive. This was despite the obligation placed on the Conseil by Article 5 of the Treaty to "take all appropriate measures . . . to ensure fulfilment of the obligations arising out of this Treaty . . ."[47]

6. THE DOCTRINE OF DIRECT EFFECT

1. REVIEW OF CASE LAW ON THE DOCTRINE OF DIRECT EFFECT

The terms "direct application" and "direct effect" are often confused and it is necessary to differentiate between these two ideas.

"Direct application" is used in the Treaty of Rome (Article 189) to denote the characteristics of regulations. Indeed, the re-enactment of provisions of regulations in national law is prohibited unless expressly or implicitly required by the regulation itself. In the case of *Variola v Italian Ministry of Finance*[48] the Court held that a regulation is immediately effective in Member States. The Court also held that Member States should not introduce any measure which might affect the jurisdiction of the Court to pronounce on any question involving the interpretation of Community law. In particular the jurisdiction of the Court under Article 177 is unaffected by any provisions of national legislation which purport to convert a rule of Community law into national law[49]. Thus direct application is the mechanism by which future Community legislation can, for the purpose of incorporation into national legal systems, be considered to have been incorporated into the Treaty itself.

The European Court of Justice has refrained from describing other secondary community legislation as "directly applicable".

In the case of *Grad v Finanzamt Traunstein*[50] the plaintiffs were long distance road hauliers who claimed that German Statute on Taxation of Road Freight Traffic was incompatible with Community rules. Before such taxation had been introduced the Commission had issued a recommendation asking that the tax not be imposed, and declaring such taxation as being incompatible with the common transport policy. It was clear that the recommendation was not directly effective as Article 189 of the Treaty expressly provides that recommendations are not binding. However, decisions of the Council of 1965 required that as soon as a common tax system had been determined by the Council this was to apply to road freight haulage. The question arose as to whether a decision, which is addressed to a Member State, could be relied upon by an individual.

The European Court of Justice stated that "although it is true that by virtue of Article 189, regulations are directly applicable and therefore by virtue of their nature capable of producing direct effect, it does not follow from this that other categories of legal measures mentioned in that article can never produce similar effects. . ." "Although the effects of a decision may not be identical with those of a provision contained in a regulation, this difference does not exclude the possibility that the end result, namely the right of the individual to invoke the measure before the courts, may be the same as that of a directly applicable provision of a regulation."

Unlike the term "directly applicable", the concept of direct effect is nowhere expressed in the Community treaties but has developed through case law of the European Court of Justice.

The idea that Community provisions can create rights and obligations that individuals may rely upon before their own national courts was further developed as a result of one of the first questions put to the European Court of Justice, and concerned a Treaty Article.

The reference came in the case of *Van Gend en Loos v Nederlandse Administratie der Belastingen*[51] in which the plaintiff sought to establish that there had been an illegal increase in customs duty. It was argved that this increase was contrary to Article 12 of the Treaty, which prohibits the introduction of new custom duties or the increase of existing ones as between Member States. Dutch, German and Belgian governments all submitted observations to the effect that alleged breaches of the Treaty by Member States were a matter to be resolved at the level of the States. They also argued that sanctions available to the Commission and Member States under Articles 169 and 170 negated the existence of individual rights.

The European Court of Justice ruled that "A restriction of the guarantees against an infringement of article 12 by member states to the procedures under articles 169 and 170 would remove all direct legal protection of the individual rights of their nationals. There is the risk that recourse to the procedure under these Articles would be ineffective if it were to occur after the implementation of a national decision taken contrary to the provisions of the Treaty. The vigilance of individuals concerned to protect their rights amounts to an effective supervision in addition to the supervision entrusted by articles 169 and 170 to the diligence of the Commission and of the member states." The European Court went on to refer to a new legal order concluding that "Independently of the legislation of member states, Community law therefore not only imposes obligations on individuals but is also intended to confer upon them rights which become part of their legal heritage."[52]

The task assigned to the European Court of Justice under Article 177 of the Treaty of Rome, the object of which is to have uniform interpretation of the treaty by national courts, confirms that the signatory States have acknowledged that community law has an authority which can be invoked by their nationals before such courts. The concept of direct effect of directives is slightly more complex as this must be regarded as a residual effect only arising out of incomplete or ineffectual implementation of such requirements, and only after the time limit for such implementation has expired. It is clear that if a directive is fully and adequately implemented by the required deadline the issue of direct effect would not arise. However, the fact that a Member State has not implemented a directive does not necessarily remove the right of individual citizens to rely on the provisions of that directive.

This point was illustrated in case *148/78 Pubblico Ministero v Ratti*[53]. The board of directors of the undertaking Silvam, of Senago (Milan), represented by Mr Ratti, decided to package its solvents and to affix to the container labels that conformed to Council Directive No. 73/173/EEC of 4 June 1973. It also decided that Council Directive No. 77/728/EEC of 7 November 1977 would be applied to its varnishes. These two directives had not been incorporated into the Italian legal system. Law No. 245 of March 1963, applying to solvents and varnishes, remained in force in Italy and was in some respects more stringent than the aforesaid directives although in others less stringent. Mr Ratti was prosecuted by the Pubblico Ministero for an infringement of Law No. 245 before the Fifth Criminal Chamber of the Pretura, Milan. However, the European Court of Justice ruled that an individual could not be held to have broken national law, which was contrary to European Community law, after the implementation date for the EC requirements had passed. This was so provided that the individual had complied with

EC legal requirements and that those requirements were "clear precise and unconditional." In other words the EC requirements could be relied upon directly by the individual. It is not necessary for the entire directive to be clear, precise and unconditional; it is sufficient if the provision to be relied upon has those characteristics. The acts complained of must also sufficiently affect the individual so as to give him *locus standi*. On the question of the second directive, for which the implementation date had not yet passed, it was held that the issue of direct effect did not apply until after such time as implementation was required.

The question of State liability was first raised in cases *C-6/90 and C-9/90, Francovich v Italian Republic*[54]. The question arose because of a failure by the Italian Government to implement Directive 80/987 on the Protection of Employees in the event of their Employer's Insolvency. The Directive provided that in the event of an employer becoming insolvent the Member State should "take the measures necessary to ensure that guarantee institutions guarantee . . . payment of employees' outstanding claims resulting from contracts of employment or employment relationships and relating to pay for a period prior to a given date" (Article 3(1)). States were free to limit the liability of guarantee institutions subject to a minimum payment of three months outstanding claims. The Italian Government took no steps to implement the Directive and in 1989 the European Court of Justice ruled that the Italian Government had failed to fulfil its obligations with respect to the requirements of this Directive. In 1991 employers of Mr Francovich became insolvent, but provisions of the Directive, which would have compensated Mr Francovich, had still not been implemented. Mr Francovich sued his former employer and the State. Mr Bonifaci, and 33 similarly aggrieved employees of another company, also sued the State.

The Italian courts dealing with the two cases referred identical questions to the European Court of Justice concerning the State's liability. The first question concerned the direct effect of the Directive and the second concerned the State's liability for damages arising out of its failure to implement the Directive. On the first question the European Court of Justice found that the Directive was not directly effective. It ruled that, although the beneficiaries of the guarantee and the size of their right (a minimum of three months outstanding pay) were sufficiently clear and precise to give direct effect, the identity of the guarantee institution was not. The Member State was not required to take on the role of the guarantee institution and as such could not be held to be responsible in this case for compensation. However, the ruling on the State's liability for damages was particularly significant.

The Court held that "where, as in the present case, a Member State fails to fulfil an obligation imposed upon it by Article 189(3) of the Treaty to take all the necessary steps to achieve the result required by the Directive, that provision of Community law, to be fully effective, must give rise to liability for damages provided that three conditions are fulfilled:

1. the results required by the Directive must include the conferring of rights for the benefit of individuals;
2. the content of these rights must be determinable by reference to the provisions of the Directive; and
3. there must be a causal link between the breach of the obligation of the State and the damage suffered by the person affected."

This decision was certainly consistent with the Court's case law. As described above it had been previously established that Community law could give rise to individual rights. It was a small step therefore to conclude that, in order to guarantee those rights, the State could be held to be liable for damages arising from any of their actions or inactions which gave rise to an infringement of those rights.

Steiner[55] concludes in her review of the case that "The breakthrough in Francovich lay not so much in the fact that the individuals were held to be entitled to claim damages against the State . . . but that their claim to compensation was independent of the principle of direct effect." Steiner goes on to suggest that "Provided the criteria for Francovich are met the individual may now proceed against the State for breach, not of substantive directly effective Community provisions, but for the State's primary failure to comply with its obligation to implement Community law." Steiner suggests that the Court could have found without difficulty that the provisions of the Directive were sufficiently clear, precise and unconditional to have direct effect. She suggests, however, that the Court deliberately failed to do so in order to establish a remedy against a Member State's infringement that did not depend on the need to prove direct effect. The prospect of liability to all parties suffering damage because of their failure to implement Community law would clearly provide States with a powerful incentive to comply fully with their Community obligations.

Although the Francovich case concerned the failure to implement a directive, the ruling was couched in very broad terms, a Member State being held to be "liable to make good damages to individuals caused by a breach of Community law for which it is responsible." Advocate General Mischo concluded that this ruling was couched in such broad terms as to be capable of being extended to a Member State's failure to observe all areas of Community law, the Treaty, Regulations or Directives, whether they are directly effective or not[56].

One issue that remained unresolved in the Francovich case was that of the level of compensation that a Member State could be expected to make. In the absence of European Community rules on this matter the level of compensation must remain a matter for the national courts to decide.

An indication of the likely view of the UK courts on such matters can, however, be seen in the case of *Bourgoin SA v Ministry of Agriculture Fisheries and Food*[57]. In this case the United Kingdom Ministry of Agriculture, Fisheries and Food (MAFF) imposed a restriction on the import of French turkey at Christmas and were sued by the French producers. In a parallel case brought by the Commission under Article 169 of the Treaty, the European Court of Justice held that measures taken by MAFF constituted a restriction on trade and as a result MAFF amended its system of import controls. By this time the turkey trade for Christmas 1982 had been affected and the plaintiffs sued MAFF for £19 million in lost trade. In this case the United Kingdom Court of Appeal accepted that there could be tortious liability for misfeasance in public office. The measure of damages would, therefore, arguably be of a tortious standard and be of a level sufficient to restore the plaintiff to a position that he or she would have occupied if the tort had not been committed.

The question remained as to whether, in the case of a referral of a question on a point of law to the European Court, effective interim relief could be obtained from the national courts. This issue is of clear significance to all who seek to rely on the doctrine of direct effect and was the essential feature of the reference by the House of Lords in the case of *Factortame Ltd v Secretary of State for Transport*[58].

This case concerned the registration of Spanish fishing vessels by English companies in order to allow them to take advantage of the United Kingdom fish quota allocation. Under the Merchant Shipping Act 1988 such ships lost their registration due to the introduction of a requirement for British citizenship together with domicile requirements. The owners of the vessels challenged these requirements on the grounds that they infringed rights given by the EEC Treaty and the common fisheries policy. It was estimated that a preliminary ruling by the European Court of Justice could have taken up to two years and the question placed before the House of Lords was whether the applicants were entitled to interim relief. The applicants argued that unless some form of interim protection were to be given they would suffer irreparable damage, with the operators of the vessels having to sell their ships and lose their livelihoods.

The House of Lords held that there was a presumption that an Act of Parliament was compatible with European Law until such time that it was declared to be incompatible and had no power to grant an interim injunction against the Crown. As such the postponement of national law was not possible. However, it was argued on behalf of the applicants that there was a fundamental requirement on national courts to ensure interim protection of rights which were bestowed as a result of direct effect of Community law. The House of Lords subsequently referred the question to the European Court as to whether Community Law obliges national courts to grant interim relief under such circumstances. In its ruling the European Court of Justice held that, "if a national court considered that the only obstacle which precluded it from granting such relief was a rule of national law it had to set that rule aside." On reference back to the House of Lords it was held that "a national court should not restrain a public authority from enforcing the law unless it was satisfied that the challenge to its validity was sufficiently firmly based to justify the exceptional course being taken."[59] In this particular case it was ruled that interim relief should be granted.

In the English case of *Twyford Parish Council and others v Secretary of State for the Environment and another*[60], the question was raised as to who could rely on the direct effect doctrine, particulary with regard to Directive 85/337 on environmental impact assessments. This case concerned the construction of a motorway near Winchester for which no environmental impact assessment had been prepared. The judge in the case concluded that there was little doubt that the applicants (Twyford Parish Council) were among those whom the Directive was intended to benefit. He also regarded the requirements of the Directive to be sufficiently clear, precise and unconditional. But, he said, the question arises whether this alone would entitle them to rely on the doctrine of direct effect in so far as the Directive had not been implemented in the prescribed time (the Directive was implemented in England and Wales on 15th July 1988, twelve days later than required). On referral the High Court held that there are no grounds for the view that, if the terms of a directive had been breached, an individual who had not suffered as a result of that breach could enforce it against the defaulting Member State.

It is interesting to note at this point that Council Directive on civil liability for damage caused by waste[61] makes specific provision for "interest groups" to seek remedy under the Directive. In the case of this Directive, this may overcome the argument that action can only be sought by injured parties. However, for those directives that do not make such provision, it does tend to support the view taken by the UK High Court.

However, Jans[62] takes a contrary view. He firstly argues that the Member State must create a means of recourse to the courts where a specific directive expressly requires it. Jans uses the example of the Directive 90/313 on Freedom of Access to Information on the Environment. Article 3 of this Directive gives any person a right to access environmental information, and Article 4 provides for a judicial or administrative review where this right is not allowed. Jans also uses, as an example, the Directive on environmental impact assessment in which the public are to be given the right to express concern, and information on environmental impact of certain projects is to be given to the public.

Jans secondly considers the question of to whom the Member State must offer a means of protecting their rights where a directive does not contain any express provision on the matter. In two joined cases concerning *Verholen v Sociale Verzekeringsbank*[63], the European Court of Justice held that the right to rely on a directive was not limited to those individuals to whom the provisions were immediately directed. Indeed, the Court said that it could not be ruled out that other persons might also have a direct interest. In the light of these decisions Jans concludes that "the decision of the High Court in the environmental impact assessment case . . . must be rejected. What is important is not whether loss has been suffered but whether an individual has a direct interest in compliance by the Member State with obligations imposed by a Directive."

Where a directive itself shows its scope, with regard to its applicability to certain groups, there is little difficulty in determining the extent of "direct effect". However, if, as in the case of the Dangerous Substances Directive, the obligation placed on Member States relates to the imposition of emission standards or the attainment of environmental quality objectives, the question arises as to whom the interested parties are.

The case of *Commission v Germany*[64] concerned the implementation of Directive 80/779 on air quality standards for sulphur dioxide in West Germany and addresses this issue. Under this Directive Member States are required to ensure that atmospheric levels of sulphur dioxide and particulates comply, as a minimum, with those set in the Directive. In Germany these limits were converted by the means of the Technische Anleitung Luft[65], a legal instrument whose mandatory nature was disputed. In this case the European Court of Justice considered Article 1 of the Directive which states that the limit values are established to improve "the protection of human health" and "protection of the environment". The Court concluded that the limits in the Directive were intended to confer general protection and as such breach of these limits conferred a general right on citizens of the Member States.

The importance of this ruling in its application to the Dangerous Substances Directive is clear. The Dangerous Substances Directive also imposes emission standards and water quality standards and it would appear, therefore, that a general right, with regard to water quality, is conferred on all the citizens of the United Kingdom and not only on those individuals who may be immediately affected.

2. REVIEW OF CASE LAW ON WHAT CONSTITUTES "ORGANS OF THE STATE".

Following the case of *Marshall v Southampton Area Health Authority*[66] the widely expressed legal opinion was that provisions of directives could only be relied upon against the "State". Miss Marshall was a senior dietician with the Southampton Area Health Authority and was compulsorily retired at the age of sixty although male employees were allowed to continue until the age of sixty-five. The Sex Discrimination Act 1975 prohibits the discrimination by an employer against an employee on the grounds of sex but this provision is excluded in relation to retirement. The Industrial Tribunal held that the dismissal of Miss Marshall did violate the principle of equal treatment but the Employment Appeals Tribunal held that violation of the terms of the Equal Treatment Directive could not be relied upon. The Court of Appeal referred to the European Court of Justice the question of whether the act of dismissal was illegal, and whether the Directive had direct effect. The European Court of Justice ruled that the dismissal was illegal and that Miss Marshall could rely directly on the provisions of the Directive.

The Court concluded:

> "It follows that a Directive may not of itself impose obligations on an individual and that a provision of a Directive may not be relied upon as such against such a person. It must therefore be examined whether, in this case, the respondent must be regarded as having acted as an individual. In that respect it must be pointed out that where a person involved in legal proceedings is able to rely on a Directive against the State he may do so regardless of the capacity in which the latter is acting, whether as employer or public authority. In either case it is necessary to prevent the State from taking advantage of its own failure to comply with Community law."[67]

In this case the Health Authority was held to be a public authority and thus part of the State; consequently Miss Marshall was successful in her appeal.

This concept of the State and "emanations" of the State was expanded further in such cases as *Johnson v Chief Constable of the R.U.C.*[68] where the Chief Constable was also regarded as "the State". In this case it was said that a public authority charged with maintaining public order does not act as an individual.

Further clarification came in the case of *Foster v British Gas*[69]. In this case six women had been compulsorily retired at the age of sixty. They claimed that by allowing differential compulsory retirement ages for men and women, the United Kingdom was in breach of its obligations to implement the Equal Treatment Directive. For precedent, as set out in the case of *Marshall v Southampton Area Health Authority*, to apply, it was necessary to demonstrate that British Gas was indeed an emanation of the State. The Court of Appeal held that an emanation of the State included an independent public authority charged by the State with the performance of any of its classic duties. Upon appeal to the House of Lords it was held that it was for the European Court of Justice to decide the categories of persons against whom the provisions of a directive may be relied upon and for the national courts to decide whether a party to proceedings before them falls within such a category. It was held that "A body, whatever its legal form, which has been made responsible pursuant to a measure adopted by the State for providing a public service under the control of the State, and has for that purpose special powers beyond those which result from the normal rules applicable in relations between individuals", would fall within such a category. This would include a commercial undertaking providing a public service if it was "under the control of the State" and was equipped with "special powers for that purpose".

In a later case of *Doughty v Rolls Royce plc*[70] Mrs Doughty was a clerk who was compulsorily retired at the age of sixty by Rolls-Royce. In this case the Court of Appeal held that for direct effect to apply all the criteria in the Foster case must be met. That is to say that the organisation must:

1. have been made responsible for providing a public service; and
2. the public service must be under the control of the State; and
3. for that purpose the organisation must have special powers beyond those that result from the normal rules applicable in relations between individuals.

In the case in question it was insufficient that the organisation (Rolls Royce) was wholly owned by the State and the doctrine of direct effect did not apply.

The rights of individuals under directly effective European law would appear to have been further strengthened as a result of the case *Marleasing S.A. v La Comercial Internacional de Alimentacions S.A.*[71]. This case concerned a Spanish company which had allegedly been formed to defraud, by the transfer of funds to a new company, the creditors of its founders. The decision of the European Court of Justice required that a Spanish national court interpret national law in accordance with the provision of a non-implemented directive giving a so called "horizontal effect". It was reasoned by the European Court of Justice that Member States are obliged by Article 5 to take all general or particular measures to ensure fulfilment of Treaty obligations. This duty applies to all authorities of Member States, including the courts. When applying national law, whether this was made before or after a directive, national courts are obliged to interpret it as far as possible in conformity with the directive. Thus, because of the duty placed on national courts to interpret national law in the light of directives, individuals can rely on directive provisions against other individuals via these courts (the horizontal effect). This interpretation means that the previous limitation on the reliance of direct effect against "organs of the State" has perhaps been removed.

After the Marleasing case judges should have regard to all directives in disputes between private persons and companies and whenever possible interpret national law such that it satisfies the aims of the directives. However, it has been reported[72] that the English Courts have been reluctant to construe existing domestic law against its *prima facie* meaning so as to give effect to a later directive.

Geddes states[73] that until recently the position in UK law appeared to be that set out in the judgement of Warner in *National Smokeless Fuels Ltd v CIR*[74], as approved by the Court of Appeal in *Rothschild Holdings plc v CIR*[75].

In the case of National Smokeless Fuels, it was decided that, subject to one exception, if the wording of statute was clear and unambiguous the terms of a directive could not be invoked to modify the meaning. The one exception to this case was where a directive requirement was clear, precise and unconditional, such as to confer a direct effect.

Geddes concludes that the position was modified by two decisions of the House of Lords; *Pickstone v Freeman plc*[76] and *Lister v Forth Dry Dock & Engineering Co. Ltd*[77].

The Lister case concerned the relations between the Business Transfer Directive (77/187) and the Transfer of Undertakings (Protection of Employment) Regulations 1981 (SI 1981/1798). This statutory instrument was issued with the sole purpose of implementing the requirements of the Directive and the dispute arose over the rights of one employee who had been dismissed one hour before the transfer of ownership of a business. If one were to consider the statutory instrument the persons concerned were not capable of being considered as employees at the moment of transfer, but if one were to look at the provisions of the Directive they were. Both employers before and after the transfers were private corporations so that the issue of "vertical effect" did not apply (vertical effect being the ability of individuals to rely on directive provisions as against the State or its emanations).

The House of Lords held that, because the statutory instrument was based on the Directive, the terms of the instrument should be construed purposively so as to conform to the Directive. Lord Oliver stated "The approach to the construction of primary and subordinate legislation enacted to give effect to the UK's obligation under the EEC Treaty has been the subject matter of recent authority in this House and is not in doubt. If the legislation can reasonably be construed to conform with these obligations, obligations which are to be ascertained not only from the wording of the relevant directives but from the interpretation placed upon it by the Court of Justice of the European Communities, such a purposive construction will be applied even though, perhaps, it may involve some departure from the strict and literal application of the words which the legislature has elected to use."[78] It follows that if the United Kingdom's national courts take the view that implementing legislation is to be given a meaning such as to meet directive obligations, even where the legislation clearly conflicts with the directive, the doctrine of direct effect will in these cases be made redundant.

The situation is different in the case of national legislation that pre-dates directives. If a directive has direct effect there will be a remedy via the national courts as described previously. However, where there is no direct effect the courts have already refused to distort the meaning of existing statute so as to give effect to a directive.

In the case of *Duke v GEC Reliance*[79] Mrs Duke was an employee of GEC Reliance and was compulsorily retired at the age of sixty, whereas male colleagues were not retired until the age of sixty-five. Mrs Duke complained that such discrimination on the grounds of sex was contrary to the Sex Discrimination Act 1975 and argued that the Act should be construed so as to give effect to the Equal Treatment Directive 26/207 dated 9 February 1989. The court in this case rejected this argument saying that "Of course a British Court will always be willing and anxious to conclude that United Kingdom law is consistent with Community law. Where an Act is passed for the purpose of giving effect to an obligation imposed by a Directive or other instrument, a British court will seldom encounter difficulty in concluding that the language of the Act is effective for the intended purpose. But the construction of a British Act of Parliament is a matter of judgement to be determined by British courts and to be derived from the language of the legislation considered in the light of circumstances prevailing at the date of the enactment."

A more recent case which effectively reiterates this line was that of *Webb v EMO Air Cargo (UK) Ltd*[80]. In this case an employee of EMO Cargo became pregnant, and prepared to go on maternity leave. EMO Cargo therefore employed Ms Webb to fill the vacancy which would subsequently become available. The employment of Ms Webb commenced six months before the commencement of the original employee's maternity leave but within two weeks of starting her employment Ms Webb informed her employer that she too was pregnant and would wish to be on maternity leave at the same time as the employee for whom she had been intended to substitute. On learning this EMO Cargo dismissed Ms Webb on the grounds that she would not be available to do the job for which she had been appointed. Ms Webb took action against EMO Cargo for unfair dismissal and sex discrimination on the grounds of pregnancy.

The action was dismissed by the Industrial Tribunal and the Employment Appeal Tribunal. On further appeal the court held that under the Employment Protection Acts and the Sex Discrimination Act 1975 the facts of the case negated discrimination. But under the Sex Discrimination Directive 76/207, as interpreted by the European Court of Justice in the case of *Dekker v Vjv-Centrum*[81], there was uncertainty whether pregnancy could ever be a justifiable ground for dismissal.

Lord Justice Glidewell was of the opinion that under the Directive it depended on the circumstances and that the circumstances of the case (the inability of Ms Webb to fulfil the post for which she had precisely been engaged) justified dismissal.

Lord Justice Balcombe felt that pregnancy was *per se* an inadmissible ground for dismissal.

The opinion of Lord Justice Bedlam was particularly interesting however. He felt that the European Court of Justice would probably support Ms Webb but that in any case the Directive did not have direct horizontal effect and that the requirement in the Marleasing case was that a court should interpret English law so as to give effect to the Directive only if such interpretation did not distort the meaning of English law. In this case, he concluded, to hold in favour of Ms Webb that pregnancy could never be grounds for dismissal in any circumstances would distort the Sex Discrimination Act, and dismissed the appeal.

The conclusion to be drawn from these rulings seems to be that where legislation has been specifically passed to give effect to directives, national courts can be expected to read into the legislation meaning that is consistent with directive obligations. However, where the statute in question predates the directive, the courts will not ignore parts of the legislative instrument if to do so would result in a distortion of the general legal meaning.

This could prove problematical in view of the United Kingdom's practice of "tacking on" directive requirements to extant legislation by the use of statutory instruments. This situation is specifically discussed below with regard to the Dangerous Substances Directive.

The issue of the Webb case was finally resolved by a recent ruling given by the European Court of Justice. The Court held that, "The termination of Ms Webb's employment contract, which was for an indefinite period, on the grounds of her pregnancy could not be justified by the fact that she was prevented, on a purely temporary basis, from performing the work for which she had been engaged."[82] The issue of the interpretation of legislation which predates a directive remains, therefore, an area of contention.

Unfortunately, the whole area of direct effect has again been confused by the case of *Wychavon District Council v Secretary of State for the Environment and Velcourt Ltd*[83]. This case concerned a planning application for the construction of poultry houses and dwellings for agricultural workers. Regulations implementing the 1985 European Community Directive on Environmental Impact Assessment, came into force in England and Wales on 15 July 1988, although the Directive itself should have been brought into force on 3 July 1988. The Velcourt application was made on 11 July and so appeared to have escaped the environmental assessment requirements under national law. The gap between the required and the actual implementation dates gave rise to the question whether the direct effect doctrine could be applied.

Vertical effect applies to emanations of the State, a description that would clearly apply to the Secretary of State and his planning officer. The developers were private parties and Mr Justice Tucker, the presiding High Court judge, held that the provisions of the Directive could not be applied to such individuals, as this would have given horizontal effect to the Directive obligations. This is in itself surprising in the light of the ruling in the case of *Marleasing* in which it was held that the Court, as an emanation of the State, should interpret national law having regard to obligations imposed by directives.

Nevertheless, the critical obligations under the 1985 Directive are placed on the planning authority and not on the private developer, and it may have been argued that the planning authority could have used its own powers and resources to require further information from the developer. Thus, Mr Justice Tucker then went on to consider whether the Directive was sufficiently clear, precise and unconditional so as to have direct (vertical) effect. He held that "it is unnecessary for me to analyse each Article in the Directive in turn in order to determine whether it is unconditional or uncertain. It will suffice if in respect of any Article which offends against the principle it is identified." He then went on to identify a number of Articles that were insufficiently clear and precise[84].

This view is clearly at odds with the previously described case concerning this Directive, *Twyford Parish Council and others v Secretary of State for the Environment and another*, where Justice McCullough ruled, "I have no doubt that the applicants were amongst those whom the Directive was intended to benefit and that its conditions were unconditional and sufficiently precise." This case again serves to illustrate the reluctance of the courts to apply a sympathetic interpretation of national law and thereby give effect to directive obligations.

The uncompromising attitude of the Secretary of State for the Environment toward the question of direct effect had been seen in an appeal, by National Smokeless Fuels Ltd, against a refusal of planning permission for a gas turbine at their Monkton works[85]. The essential features of this case are that the planning authority refused permission for National Smokeless Fuels to build a power generation plant using gas from coke ovens. The local planning authority urged the Secretary of State, when considering the appeal, to link any permission to a requirement for the desulphurization of the coke oven gas. Her Majesty's Inspectorate of Pollution had, however, already registered the plant accepting that "best practicable means" for the prevention of noxious emissions comprised the use of a 55 metre high chimney.

The Counsel for the regional action group (who were opposed to the development) argued that the development constituted a substantial change to the existing Monkton plant, requiring the works to obtain prior authorization under the provisions of Article 3(2) of the Directive concerning the combating of air pollution from industrial plants[86]. Article 4 of this Directive requires that an authorization may only be issued when the competent authority is satisfied that:

"1. all appropriate preventative measures against air pollution have been taken, including the application of the best available technology, provided that the application of such measures does not entail excessive costs;

2. the use of the plant will not cause significant air pollution particularly from the emission of substances referred to in Annex II; [to that directive]

3. none of the emission limit values applicable will be exceeded;

4. all the air quality limit values applicable will be taken into account."

The Counsel for the resident's action group argued that the proposals of the application failed to meet the objectives of the Directive, that is to say the prevention and reduction of air pollution. It was also argued that if, nevertheless, the Secretary of State were to grant permission, measures to ensure a substantial reduction of the polluting potential of the proposed plant should be incorporated as a planning condition.

The Secretary of State dismissed this argument, relying on the expert opinion of HMIP that the 55 metre high chimney constituted the best practicable means for dealing with the noxious releases. More telling, however, was his response on the issue of direct effect. He said "Directive 84/360/EEC applies to coke ovens but, as the Commission of the [EEC] has been formally notified, its provisions are presently implemented by the Alkali, etc. Works Regulation Act 1906, the Health and Safety at Works etc. Act 1974 and Regulations made under these Acts."[87]

It may be inferred from this that it is the Secretary of State's view that, unless the Commission challenges the adequacy of national implementing legislation, the doctrine of direct effect has no relevance.

7. THE APPLICABILITY OF THE DOCTRINE OF DIRECT EFFECT TO REGULATORS OF DANGEROUS SUBSTANCES IN ENGLAND AND WALES

It is first necessary to examine the three regulatory bodies involved with the control of dangerous substances.

1. THE PRIVATE WATER COMPANIES

Before 1989 all aspects of the water cycle were overseen and regulated by ten regional water authorities. The water authorities were closely regulated by the Department of the Environment, although they were not part of any government department. In 1989 the water industry was separated into its operational functions and its regulatory functions, the former being sold by the government to private investors.

The privatised water companies have a limited role in the control of dangerous substances, although a residual role does still exist. In the case of processes which are not controlled under the provisions of the Environmental Protection Act 1990, releases of List I substances to sewer are authorized by the Secretary of State using the referral procedure set out in Sections 120 and 130 of the Water Industry Act 1991.

Under the provision of the Water Industry Act 1991 and the associated prescribed substances regulations[88], applications for the proposed release of List I substances to sewer have to be referred to the Secretary of State for authorization. This obligation also applies to proposed variations of extant consents by the water companies, where such variations would authorize a release of one or more List I substances above the background concentration. Background is defined in Statutory Instrument 1989 No.1156[89] as "such concentration of the substance as would, but for anything done on the premises in question, be present in the effluent discharged from those premises; and without prejudice to the generality of the foregoing, includes such concentrations of the substance as are present -

a. in water supplied to the premises;
b. water abstracted for use in the premises; and
c. in precipitation onto the site within which the premises are situated."

Through the referral procedure, the Secretary of State sets discharge limits for the prescribed substances. These limits are communicated to the water companies (in practice to the sewerage undertaker, which is one part of the water company) in a Notice of determination. The water companies are then required to use their powers to issue, vary, or refuse consent, in order to give effect to the requirements laid down by the Secretary of State in that Notice.

It is generally accepted that the water companies are at liberty (subject to appeal by the discharger) to set standards that are more stringent than those that are required by the Secretary of State. Clearly, if a discharger is required to comply with more stringent standards, compliance with the Secretary of State's limits will also be assured. The water company concerned has, therefore, fulfilled its obligations. This ability to impose more onerous standards places the water companies in an interesting legal position. In these circumstances the release has been authorized by the competent authority (the Secretary of State) but, for practical purposes, the actual limits imposed are those set by the water company. This is of particular relevance to authorization of new or substantially increased discharges. In these circumstances the dangerous substances daughter directives require that authorization is only granted where the release has been minimised using the best technical means available. The legal certainty given by these provisions is derogated by the ability in national law for the sewerage undertakers to impose more stringent standards. What is not clear is whether Article 10 of 76/464, giving Member States the ability to impose more stringent measures, is to be applied by the State in its widest sense, i.e. nationally or whether emanations of the State can choose to impose more stringent measures on a site by site basis. This is discussed further below.

A similar situation exists where the process gives rise to a discharge to sewer but is also controlled by HMIP under the provisions of the Environmental Protection Act 1990. In this case provisions contained within the Water Industry Act 1991 for the Secretary of State to control List I substances are removed[90]. Instead, the discharge of the List I substance is authorized by the competent authority by the granting of an authorization under Section 6 of the Environmental Protection Act 1990. Nevertheless, the operator of the process will still require a consent from the water company before a discharge of trade effluent can be made to sewer, this consent operating in parallel to the authorization. Again, the water company is at liberty to impose tighter standards than those embodied within the HMIP authorization. HMIP is required to have regard to cost in their derivation of limits whereas water companies are not. It is almost inevitable, under these circumstances, that discharge limits set by the two different regulators will differ. It can be seen that here again it is the water companies who are setting the actual standards for List I substances.

It remains to be considered whether the private water companies are to be regarded as emanations of the State and on this point the previously discussed case of *Foster v British Gas* is particularly relevant. To restate the decision of the European Court of Justice, the provisions of a directive can be relied upon against organisations or bodies which are:

1. subject to the authority or control of the State; and
2. have powers beyond those which result from the normal rules applicable to relations between individuals.

This includes a body, whatever its legal form, which has been made responsible, pursuant to a measure adopted by the State, for providing a public service under the control of the State and has for that purpose special powers beyond those which result from the normal rules applicable in relations between individuals.

We now examine each of these points. Firstly, the private water companies are clearly providing a public service. Section 37(1) of the Water Industry Act 1991 sets out the duty of a water company as a water undertaker "To develop and maintain an efficient and economical system of water supply within its area and to ensure that all such arrangements have been made –

a. for providing supplies of water to premises in that area and for making such supplies available to persons who demand them; and
b. for maintaining, improving, and extending that water undertaker's water mains and other pipes, as are necessary for securing that the undertaker is and continues to be able to meet its obligations under this Part."

A parallel duty exists under section 94(1) of the Act in relation to the duty of the water companies as sewerage undertakers.

Secondly, the Companies are, to a significant extent, under the control of the State, particularly concerning the major part of their financing ie. consumer charging. The Secretary of State for the Environment, using powers under Part I Section 2 of the Water Industry Act 1991, has appointed a Director General of Water Services. The Director's duties include:

1. the protection of the interests of customers with respect to charges;
2. the protection of the interests of customers with respect to quality of service; and
3. the promotion of economy and efficiency on the part of the undertakers[91].

Finally, the water companies have special powers for the purpose of carrying out their functions, the Water Industry Act 1991 furnishing the water companies with the necessary powers to operate. Examples of such "special powers" include the power to impose temporary hose pipe bans, the power to alter drainage systems of premises within an area, the power of compulsory purchase (with the permission of the Secretary of State), the power to make bye-laws and the power to lay pipes in streets and other land.

It is concluded, therefore, that a water company may be considered as an emanation of the State and as such subject to the doctrine of direct effect.

This view was confirmed by Justice Blackburne who, in the case of *Griffin v South West Water Services Ltd*[92], examined the three conditions set out in the case of *Foster v British Gas* in so far as they related to South West Water as a water company. This case concerned the provisions of Council Directive 75/129/EEC relating to collective redundancies, which, the plaintiff claimed, were directly enforceable against South West Water Services Ltd.

Justice Blackburne held that South West Water (in common with all water companies):

1. had been made responsible for providing a public service;
2. it had been provided with special powers for the purpose of providing that service;
3. the issue is not whether the body in question is under the control of the State but whether the public service in question is under the control of the State;
4. the legal form of the body is irrelevant;
5. the fact that the body is a commercial concern is irrelevant;
6. the fact that the body is not an agent of the State is irrelevant; and
7. the fact that the State does not possess day-to-day control of the activities of the body is also irrelevant.

Justice Blackburne held that the requirements of the Directive were directly enforceable against the water company.

2. THE NATIONAL RIVERS AUTHORITY

The NRA was established at the time of privatisation under the provisions contained in the Water Act 1989. The NRA has a wide brief which, among other things, concerns the regulation of discharges of dangerous substances into controlled waters. Although the NRA is a stand-alone organisation, which is in part self-financing (also receiving grant aid from the Treasury), it meets all the criteria required in the *Foster v British Gas* test. That is, it provides a public service (environmental regulation, amenities for recreation, fisheries etc), it does so under the control of the State (control as exercised by the issuing of Directions from the Secretary of State for the Environment) and it has special powers for this purpose (The Water Resources Act 1991). The NRA must, therefore, be regarded as an emanation of the State and as such liable to the doctrine of (vertical) direct effect.

In addition to powers conferred by the Water Resources Act 1991, the NRA also has powers under Part I Section 28(3) of the Environmental Protection Act 1990. This section states that:

"(3) Where the activities comprising a prescribed process designated for central control include the release of any substances into water included in waters which are controlled waters for the purposes of Chapter I of Part III of the Water Act 1989, then-

a. the enforcing authority shall not grant an authorisation [sic] under this Part if the National Rivers Authority certifies to the enforcing authority its opinion that the release will result in or contribute to a failure to achieve any water quality objective in force under Part III of that Act; and
b. any authorisation [sic] that is granted shall, as respects such releases, include (with or without others appearing to the enforcing authority to be appropriate) such conditions as appear to the National Rivers Authority to be appropriate for the purposes of this Part as that Authority requires by notice in writing given to the enforcing authority".

Thus, just as the sewerage undertaker may impose more onerous conditions than those required by the Secretary of State, the National Rivers Authority can require the imposition of more onerous conditions in an authorization than HMIP (acting as the competent authority) would have sought to impose.

3. HER MAJESTY'S INSPECTORATE OF POLLUTION

HMIP is a part of the Department of the Environment and as such is to be regarded as a government department. The situation here is the clearest of all three of the regulatory bodies and HMIP is obviously to be considered an emanation of the State.

To summarise, therefore, the three regulatory bodies involved with the control of release of List I substances are all to be regarded as emanations of the State. The consequence of this is that private organisations can rely on obligations placed on the State by any directive provisions that are directly effective. Failure to recognise this duty may leave the regulators open to action in the national courts, resulting in possible liability for the payment of compensation. Regulators should consider this position, and view implementation of national law in the light of the European Community law that it is intended to implement.

Tromans[93] concurs with this view and states "It can be seen that many of the provisions of Pt.I of the Act [The Environmental Protection Act 1990] are framed with a view to complying with the requirements of these and other Directives [76/464/EEC and its daughters], and that those provisions will need to be read in the light of applicable E.C. law."

8. DANGEROUS SUBSTANCES

1. BACKGROUND TO THE CONTROL OF DANGEROUS SUBSTANCES

The Commission stated in its communication on a move "Toward Sustainability" that "Water is one of the elementary sources of life. Water quality is also an indicator of the general quality of the natural environment. Without water, a harmonious and sustainable maintenance or development of socio-economic activities is not possible."[94]

Water was one of the first sectors of the environment to attract the attention of the European Community and the early measures date from the Community's action plans of 1973 to 1976. As a result of subsequent action programmes covering the years 1977 to 1992, Howarth concludes that "water is now probably the environmental sector with the most coverage under Community Law."[95] The list of directives in this area is extensive and covers inland waters, ground waters, estuarine waters, water intended for drinking, bathing waters and waters intended for use as fisheries. It is among this general background of pollution control measures that we find Directive 76/464/EEC, a framework directive that establishes a mechanism for the reduction and elimination of pollution by certain dangerous substances.

2. THE DANGEROUS SUBSTANCES DIRECTIVE

The Directive on pollution caused by certain dangerous substances discharged into the aquatic environment of the Community 76/464/EEC (the so called "dangerous substances" directive), sets a framework for pollution control. Proposed in October 1974, it was notified in the Official Journal of the Community in May 1976. Unusually, Directive 76/464/EEC contains no implementation date, but the Commission suggested the following deadlines in a written communication to Member States dated 3 November 1976.

(1) A system of prior authorization for the release of dangerous substances to be in place by 15 September 1978;
(2) pollution reduction programmes for List II substances to be in place by 15 September 1981;
(3) such programmes to be implemented by 15 September 1986.

The Directive was adopted under the provisions of Article 100 of the Treaty, which concerns the "issue of directives for the approximation of such provisions laid down by law, regulation or administrative action in Member States as it directly affect the establishment or functioning of the common market."

The preamble to the Directive states "Whereas any disparity between the provisions on the discharge of certain dangerous substances into the aquatic environment already applicable or in preparation in the various Member States may create unequal conditions of competition and thus directly affect the functioning of the common market; whereas it is therefore necessary to approximate laws in this field . . . [The Council of The European Communities] has adopted this directive." Thus, although of direct relevance to environmental protection, the Directive was adopted (in the absence of more suitable provisions) on the basis of equalisation of the market.

The Directive sets a framework for the reduction or, where possible, the elimination of pollution of inland, coastal and territorial waters by certain dangerous substances. Discharges of substances set out in List I to the Directive must comply with agreed European Community-wide standards which are laid down in two distinct control regimes.

The first method, the Uniform Emission Standards approach , sets emission limits which have to be complied with irrespective of the size or number of discharges or the nature of the receiving water. All Member States, except the United Kingdom, favoured this approach which seeks to apply technology-based standards to minimise the release of dangerous substances. This approach has advantages where water courses cross national boundaries; for example, it would be difficult to apply any control on releases which permitted the Member State at the headwaters of a river to pollute the water to the detriment of the downstream Member State.

The second method of control mentioned in the Directive is based on environmental quality objectives. Article 6 of the Directive requires that limit values established by the Directive are applied except in the case where the Member State can prove to the Commission (in accordance with set procedures) that environmental objectives are being met. The Article allows for the Member State to apply more severe standards than those set out if it so wishes.

Each of these forms of control has its own advantages and disadvantages.

The setting of standards in consents to discharge, when adopting the Uniform Emission Standards approach, is simple. Such consent conditions are perceived by industry to be equitable, offering no trade advantages to different operators at different locations. Most importantly the Uniform Emission Standards approach minimises the release of dangerous substances by the use of the "best technical means available"[96]. However, (unlike the use of environmental quality objectives) this method of control does not take any account of the impact of the discharge or the cumulative impact of several discharges; nor does it provide a mechanism for controlling diffuse inputs.

The Dangerous Substances Directive places obligations on Member States to control substances or groups of substances as set out in two lists annexed to the Directive.

List I contains certain individual substances which belong to families and groups of substances selected mainly on the basis of their toxicity, persistence and bioaccumulation and are specified in the annex to the Directive as:

1. organohalogen compounds and substances which may form such compounds in the aquatic environment;
2. organophosphorous compounds;

The term "Uniform Emission Standards approach" is often used in Britain. However, this term does not appear in the Dangerous Substances Directive. The Directive requires that "limit values" are set at the Community level and that Member States set emission standards, on discharges, at least as stringent as the limit values. Nevertheless, the term "Uniform Emission Standards approach", when used to contrast with the Environmental Quality Objectives approach, is in such common use that it is also adopted here to describe the application of limit values.

3. organotin compounds;
4. substances in respect of which it has been proved that they possess carcinogenic properties in or via the aquatic environment;
5. mercury and its compounds;
6. cadmium and its compounds;
7. persistent mineral oils and hydrocarbons of petroleum origin;
8. persistent synthetic substances which may float, remain in suspension or sink and which may interfere with any use of the waters.

These are the substances in respect of which pollution is to be eliminated.

The Directive also requires that pollution caused by List II substances, those substances with a lesser potential for causing environmental damage, is minimised.

List II contains:

- substances belonging to the families and groups of substances in List I for which the limit values have not been determined.

- certain individual substances and categories of substances belonging to the families and groups of substances listed below, and which have a deleterious effect on the aquatic environment, which can, however, be confined to a given area and which depend on the characteristics and location of the water into which they are discharged. These families and groups of substances are:

1. The following metalloids and their compounds;

1. Zinc	8. Antimony	15. Uranium
2. Copper	9. Molybdenum	16. Vanadium
3. Nickel	10. Titanium	17. Cobalt
4. Chromium	11. Tin	18. Thallium
5. Lead	12. Barium	19. Tellurium
6. Selenium	13. Beryllium	20. Silver
7. Arsenic	14. Boron	

2. Biocides and their derivatives not appearing in List I.
3. Substances which have a deleterious effect on the taste and/or smell of the products for human consumption derived from the aquatic environment, and compounds liable to give rise to such substances in water.
4. Toxic or persistent organic compounds of silicon, and substances which may give rise to such compounds in water, excluding those which are biologically harmless or are rapidly converted in water into harmless substances.

When 76/464/EEC was adopted, it was envisaged that proposals for "daughter" directives, which would set standards for an initial set of List I substances, would be made by the Commission within two years. Unfortunately, due mainly to dissent between the United Kingdom and the rest of the Community concerning procedures for identifying priority substances, it was not until 1982 that the first daughter directive (dealing with mercury discharges from chlor-alkali plants) was adopted. This was followed in 1983 by the daughter directive on cadmium discharges (83/513/EEC), and in 1984 by the daughter directives on mercury discharges from industrial sectors other than the chlor-alkali industry (84/156/EEC) and on discharges of hexachlorocyclohexane (84/491/EEC).

Each time that such a directive was adopted the essentially standard preamble and text needed to be agreed by all the Member States. In 1986, in order to try to speed up the adoption procedure, daughter Directive 86/280/EEC was adopted. This contains a set of standard clauses and an annex which contains limit values and environmental quality objectives for individual List I substances. The substances for which standards are laid down in 86/280/EEC are carbon tetrachloride, DDT and pentachlorophenol. This Directive is often referred to as the "standard articles" Directive and represents a "fast track" procedure for introducing standards for further List I substances. This so called "fast track" approach was used when the annex to this Directive was amended in 1988 by Directive 88/347/EEC, to include within List I aldrin, dieldrin, endrin, isodrin, hexachlorobutadiene and chloroform. Further additions were made in 1990 by 90/415/EEC which requires control of perchloroethylene, trichloroethylene, 1,2 dichloroethane and trichlorobenzene.

It should be pointed out that the limit values imposed by all of these daughter directives only apply to the industry sector named within the directive. For example, the cadmium Directive sets emission standards for "Zinc mining, lead and zinc refining, cadmium metal and non-ferrous metal industry, manufacture of cadmium compounds, manufacture of pigments, manufacture of stabilizers, manufacture of primary and secondary batteries, and electroplating." However, control by the Environmental Quality Objectives approach necessarily requires that all releases to water are controlled so that compliance with the objective can be ensured. Thus, some sources which are not regulated elsewhere in the Community (under the Uniform Emission Standards approach) will be controlled in the United Kingdom in order to protect quality objectives. It is also likely that while industry which is sited on large rivers or estuaries has benefited from the "dilute and disperse" philosophy that environmental quality objectives allow, other industry with less available dilution may have been disadvantaged, ie. standards imposed on inland discharges may need to be more stringent than those required by the Uniform Emission Standards approach.

There are several fundamental aspects of the framework directive and its daughters. It is necessary to consider each of these separately, and to consider how each of these obligations has been implemented in England and Wales. The following twelve sections examine these arrangements.

(1) Constraints imposed by the Dangerous Substances Directives are to be applicable to inland surface waters, territorial waters, internal coastal waters, and ground water [Article 1, 76/464/EEC]. This final aspect is now superseded by a specific directive relating to the protection of ground waters[97].

The provisions of the Water Resources Act 1991, by which all discharges to surface and ground waters are controlled (including List I substances), apply to "controlled waters".

Controlled waters are defined as:

relevant territorial waters;
coastal waters;
inland freshwaters including lakes, ponds and rivers;
and ground waters[98].

The system of control under the Water Resources Act 1991 can be seen, therefore, to deal adequately with all of the areas required under this Article. The Directive also requires that discharges of List I substances to sewer are controlled "where this is necessary for the implementation of this Directive." [Article 3, 76/464/EEC].

It has been argued by the water industry[99] that, as releases from sewage works are authorized by a competent authority (now the NRA), the United Kingdom government has set in place an unnecessary level of control in requiring that all discharges into the sewerage system are also authorized. Although this additional level of control may appear to be superfluous, Article 10 does allow Member States to take more stringent measures than those provided under the Directive. Authorization of such releases to sewer is discussed more fully below.

(2) "Appropriate steps" are to be taken to eliminate pollution from List I substances and to reduce pollution by List II substances. [Article 2, 76/464/EEC].

Article 2 sets out the objective of the elimination of pollution by List I substances and the reduction of pollution by List II substances. Unfortunately the Directive is not clear as to what is meant by "elimination" and "reduction".

"Pollution" is defined in Article 1 to the Directive as "the discharge by man, directly or indirectly, of substances or energy into the aquatic environment, the results of which are such as to cause hazards to human health, harm to living resources and to aquatic ecosystems, damage to amenities or interference with other legitimate uses of water." Because pollution is defined in terms of "harm", pollution can be considered to have been eliminated if the concentration of dangerous substances in the receiving water are such as to:

1. afford a reasonable level of protection to aquatic life; and
2. ensure that the water in question is suitable for its required use.

However, the main thrust of the Directive is in the control of discharges by ever tightening emission standards. In particular new and substantially increased discharges are to be minimised using the best technical means available. The Directive also differentiates between the reduction of pollution by List II substances and the elimination of pollution by List I substances. One may conclude, therefore, that the Directive has as its intention the complete elimination of releases of List I substances. Indeed, the Department of the Environment in its Circular 18/85[100] requires that environmental quality objectives should not be seen as rigid targets, the long term aim being to eliminate releases of List I substances. This position would, however, be more effectively achieved by the prohibition of the use of substances rather than the control of consequent releases. This approach has been used to control such substances as DDT and polychlorinated biphenyls.

(3) Discharges of List I substances require prior authorization by "the competent authority" within the Member State. [Article 3(1), 76/464/EEC]

For this obligation it is appropriate to look at the regulation of dangerous substances prior to 1989 before examining the present regulatory arrangements.

Before water privatisation, all discharges to rivers and sewers required the consent of the water authorities. In the case of discharges to rivers, consents were required under the provisions of Section 34 of the Control of Pollution Act 1974 and in the case of discharges to sewer, under Section 2 of the Public Health (Drainage of Trade Premises) Act 1937 and Section 59 of the Public Health Act 1961. Proposed consents for discharges to rivers from the authorities' own sewage works were required, under Section 35 of the Control of Pollution Act 1974, to be referred to the Secretary of State for determination. Consent was then granted, under Section 34 of the Control of Pollution Act 1974, by the Water Pollution Inspectorate (part of the Department of the Environment) on behalf of the Secretary of State.

The regulatory position changed in the autumn of 1989 when the water authorities were privatised and split into two separate bodies, responsible separately for the operational functions (such as water supply and sewage disposal), and the regulatory role. The regulatory role was given to the newly formed NRA, which was set up under the provisions of the Water Act 1989, the operational functions remaining with the private water companies.

This change in responsibilities, and the passing of the water companies into the private sector, had consequences for the regulation of dangerous substances.

Article 3 of the Dangerous Substances Directive requires that all discharges of List I substances be given prior authorization by a "competent authority". The former water authorities were deemed competent for this purpose, and the NRA is deemed competent for the purpose of controlling discharges to controlled waters. However, the Commission informed the United Kingdom government that the private water companies could not be considered as "competent" for the purpose of authorizing releases of List I substances to sewer. This decision is particularly significant if one considers the ruling of Justice Blackburne in the case of *Griffin v South West Water Services Ltd.* He concluded that a private water company meets the three conditions as set out in the case of *Foster v British Gas*. In essence, this means that a water company is to be regarded as an emanation of the State. Nevertheless, in view of the comments made by the Commission, the responsibility for authorizing releases of List I substances passed at the time of privatisation to the Secretary of State, and HMIP is required to carry out this role on his behalf. Provisions for such control are contained in Sections 120, 127, 130, 131 and 132 of the Water Industry Act 1991.

Most discharges of trade effluent to sewer require the consent of the receiving sewerage undertaker (an operational function of the privatised water companies). When control of trade effluent was introduced in the mid-1930's, several types of trade effluents were exempt. The vast majority of these exemptions have now been removed, although H.M prisons still enjoy this benefit by virtue of Crown exemption.

As described above, under the provisions of the Water Industry Act 1991 and the associated prescribed substances regulations, applications for the release of List I substances have to be referred to the Secretary of State for authorization. This obligation also applies to proposed variations of consent by the water companies where such variation would authorize a release of one or more List I substances above the background concentration. However, there are many existing consented releases of List I substances which, because no variation of consent has been made, are not yet authorized by the Secretary of State. These discharges are still made by virtue of the original trade effluent consents which were set by the water authorities and have now become vested in the privatised water companies. It is unclear whether such consents, having passed into the jurisdiction of the water companies, can still be regarded as being authorizations by a "competent" authority. Additionally, some prisons (still subject to Crown immunity) operate metal plating lines, and it is possible that a few discharges of cadmium remain unconsented and therefore unauthorized.

Finally, all of the regulatory powers given to the Secretary of State by the Water Industry Act 1991 rely on the fact that the water companies wish to control discharges by the trade effluent consenting process. If, as is the case for mercury discharges from dental surgeries, they take the view that the releases are trivial and do not insist that they apply for a trade effluent consent, the Secretary of State is left without any powers of control (other than to revoke the water companies' licences to operate!).

It is also important to note that the Water Industry Act 1991 refers exclusively to the Secretary of State. Conditions such as reporting requirements, contained in the documents issued by HMIP under provisions of the Water Industry Act, refer exclusively to HMIP but are issued by an inspector who is "authorized by the Secretary of State". This arrangement presupposes that such responsibility has been formally delegated by the Secretary of State. Despite enquiries, it has not been possible to ascertain whether any such formal transfer of powers ever took place.

There are also problems with the identification of the location of discharges of List I substances to sewer. Under the provisions of Article 3 *all* discharges liable to contain List I substances require authorization by the competent authority. There is no *de minimis* provision to take account of, for example, mercury contamination in sodium hydroxide or cadmium in zinc salts. However, if a discharge to sewer is authorized for one or more List I substances, the NRA will then authorize the discharge from the receiving sewage works for those substances. The charge which is levied on discharges from sewage works is directly related to the consented volume of the discharge and the consented substances. List I substances, particularly the organic List I substances, have a major impact on the charge, yielding increases of up to five times that incurred by a normal sewage works effluent of equal volume. The charge levied by the sewerage undertaker on trade effluent dischargers does not take account of these additional NRA charges (being based solely on organic strength, suspended matter and volume). Until such time as there is a mechanism to pass these additional costs on to the trade effluent discharger, there is a significant financial dis-incentive which discourages the water companies from identifying currently unconsented releases of List I substances to sewer.

The situation is further complicated where sewerage undertakers, on occasions, include "good housekeeping" limits for List I substances in trade effluent consents. That is to say that cautionary standards may be set in consents even where List I substances are not handled on sites giving rise to trade effluent discharges.

Unfortunately, HMIP has neither the local knowledge, nor the resource, to identify all sources of List I substances and as such it is possible that the United Kingdom is in breach of its responsibilities. This is not a major environmental issue for trace contamination of raw materials but highlights the fact that there is no mechanism in place for the identification of significant releases of new List I substances. For example, the directive on chlorinated solvents[101] requires that all releases in excess of 30kg/annum of dichloroethane, trichloroethylene, perchloroethylene and all releases of trichlorobenzene, be authorized by 1 January 1993.

When discharges of List I substances were the responsibility of the water authorities, the notification of a new directive would have set in motion actions, such as surveys or inspections by trade effluent officers, to identify releases of the relevant substances. A Department of the Environment communication, concerning the implementation of Directive 86/280/EEC relating to the control of releases of DDT, pentachlorophenol and carbon tetrachloride, states "Water authorities are asked to investigate the possibility of previously unidentified discharges from industrial sources, with particular reference to the type of plant listed under Heading A of Annex II.III to the Directive."[102] Under the present regulatory arrangements, the local knowledge remains with the water companies' trade effluent officers, but, there is no requirement for them to inform the competent authority (HMIP) of the location of such releases. Consequently, although the date for authorization has now passed, few such authorizations have been issued and the United Kingdom is again potentially in breach of its obligations under this Directive.

The table below shows the number of authorizations issued, under the provisions of the Environmental Protection Act 1990, prior to the implementation date for this directive (1 January 1993) and those issued subsequently. It is unlikely that the releases authorized during 1993 and 1994 are all new releases. It is more likely that these substances have now been authorized as a result of the determination of applications made, under the provisions of the Environmental Protection Act 1990, by the chemical industry. Some additional "authorizations" may have been granted under the provisions of the Water Industry Act 1991 but it seems likely that many sources remain unauthorized.

Authorization of List I chlorinated solvents.

	Number of Authorized Processes		
Substance	1992	1993	1994
Perchloroethylene	1	2	6
Trichloroethylene	–	2	5
1,2 Dichloroethane	–	–	8
Trichlorobenzene	–	1	3

It is also worth noting that, although discharges of List I substances from sewage works are the subject of consent from the National Rivers Authority (and therefore authorized by a competent authority), the problem of releases via storm sewage overflows has only rarely been addressed. Most sewerage systems receive varying amounts of rainwater under wet weather conditions. Sewers have historically been designed to pass flows equivalent to six times the dry weather flow forward to the receiving sewage works for treatment. All flows in excess of six times the dry weather flow are discharged via overflows on the sewers to the local watercourses. Under wet weather conditions the contents of the sewer become so diluted by rainwater that at the moment of operation of the overflow the effluent discharged is weak and, therefore, has minimal environmental impact. Unfortunately, many of the old urban sewerage systems have become progressively overloaded such that storm overflows operate before the original design wet weather flows are reached. In some case overflows operate even under dry weather conditions. It has never been common practice of the water authorities, nor is it current NRA practice to consent such releases. These potential sources of List I substances remain unauthorized.

Thus to summarise, the arrangements for the regulation of dangerous substances after privatisation can be described as:

(1) all direct discharges of List I and List II substances to inland, coastal and territorial waters required consent of the NRA (Chapter II of the Water Resources Act 1991);

(2) discharges of trade effluent to sewer required the consent of the newly privatised water companies (Section 118 Water Industry Act 1991);

(3) discharges of List I substances to sewer required the authorization of the "competent authority", in this case HMIP acting on behalf of the Secretary of State (Section 132 Water Industry Act 1991).

This situation changed further because of the introduction of the Environmental Protection Act 1990. Under these provisions the release of all substances to water (including List I and II substances) from certain processes which are prescribed for control by HMIP, require authorization from the Chief Inspector. Such an authorization provides a statutory defence under section 88(b) of the Water Resources Act 1991 and as such removes the need for a consent from the NRA for a discharge to controlled waters.

Similarly, releases to sewer from prescribed processes (defined as a release to water in Section 1(11)(c) of the Environmental Protection Act 1990) require authorization from HMIP. Such an authorization then removes the regulatory requirements of the Water Industry Act 1991[103].

The existing regulatory arrangements are shown in Figures 3–5.

Non-Prescribed processes. Figure 3.

(a) Discharges to controlled waters of all substances from non-prescribed processes (processes not prescribed, under Statutory Instrument 1991 SI472 as amended, for control under the Environmental Protection Act 1990) are consented by the National Rivers Authority.

(b) Discharges of trade effluent to sewer from such processes are consented by the sewerage undertaker except that discharges of substances, or discharges arising from processes, prescribed under Statutory Instrument 1989 SI1156 as amended, are subject to control by the Secretary of State under the Water Industry Act 1991.

(c) Discharges from the sewage works receiving trade effluent from these processes are consented by the National Rivers Authority.

(d) Liquid or solid waste disposed of by other routes is the subject of control by the Waste Regulatory Authority.

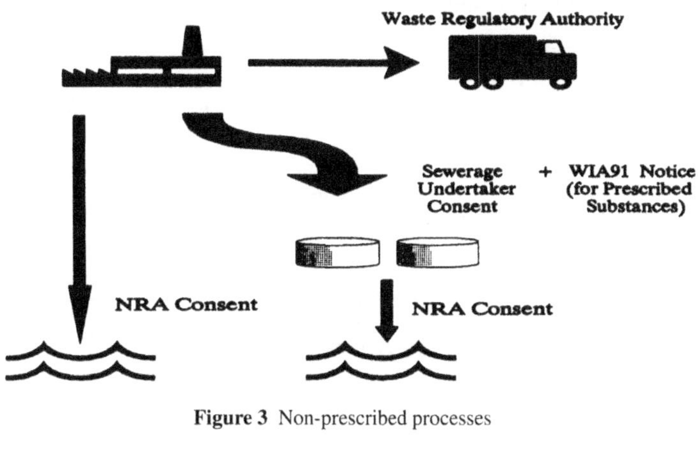

Figure 3 Non-prescribed processes

Prescribed processes (processes prescribed, under Statutory Instrument SI472 1991 as amended, for control by HMIP under the provisions of the Environmental Protection Act 1990). Figure 4.

(a) Discharges to controlled waters of all substances from prescribed processes are authorized by HMIP. The National Rivers Authority are a statutory consultee in the authorization procedure.

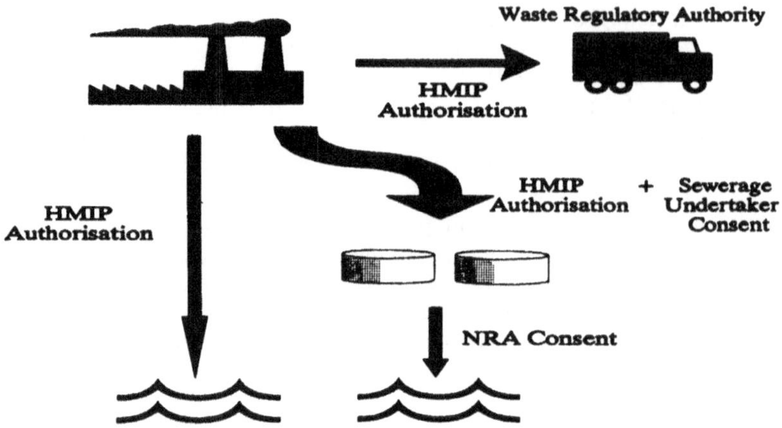

Figure 4 Prescribed Processes (processes prescribed, under Statutory Instrument SI472 1991, as amended, for control by HMIP under the provisions of the Environmental Protection Act 1990)

(b) Discharges of all substances to sewer are authorized by HMIP although consent from the sewerage undertaker is still required for all substances contained within the trade effluent discharge.

(c) Discharges from the sewage works receiving the trade effluent from these processes are consented by the National Rivers Authority.

(d) Liquid or solid waste disposed of by other routes require authorization by HMIP although the final disposal of such waste is the subject of control by the Waste Regulatory Authority.

The situation described here is one where both prescribed and non-prescribed processes are carried out on the same site. Figure 5.

(a) Discharges of trade effluent to sewer which arise from non-prescribed processes (processes not prescribed, under Statutory Instrument 1991 SI472 as amended, for control under the Environmental Protection Act 1990) are consented by the sewerage undertaker except that discharges of substances, or discharges arising from processes, prescribed under Statutory Instrument 1989 SI1156 as amended, are subject to control by the Secretary of State under the provisions of the Water Industry Act 1991.

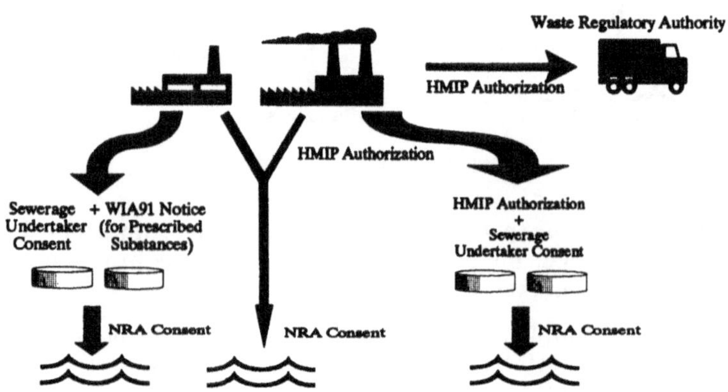

Figure 5 The situation described here is one where both prescribed and non-prescribed processes are carried out on the same site

(b) Discharges of all substances to sewer, which arise from the prescribed process (processes prescribed, under Statutory Instrument 1991 SI472 as amended, for control under the Environmental Protection Act 1990), are authorized by HMIP although consent from the sewerage undertaker is still required for all substances contained within the trade effluent discharge.

(c) Where discharges are made from both prescribed and non-prescribed processes, via a common pipe to controlled waters, HMIP authorizes the release from the prescribed process into the pipe and the NRA consent the combined discharge from the distal end of that pipe.

(d) Discharges from the sewage works receiving trade effluent from the processes are consented by the National Rivers Authority.

(e) Liquid or solid waste disposed of by other routes, from the prescribed process, require authorization by HMIP although the final disposal of such waste is the subject of control by the Waste Regulatory Authority.

(4) The emission standards laid down in the authorization shall determine "the maximum concentration of a substance permissible in a discharge" and "the maximum quantity of a substance permissible in a discharge during one or more specified periods of time." [Article 5, 76/464/EEC].

As a result of experience gained during the regulatory tenure of the regional water authorities, and following the instructions set out in Circulars 18/85 and 7/89, the practice of setting limits on concentrations and loads has become common practice in the NRA and the water companies. However, there is limited experience of such matters within HMIP. There is, therefore, potential for inconsistency in the way in which HMIP controls those discharges that fall subject to the provisions of the Water Industry Act 1991 and the Environmental Protection Act 1990. The provisions of regulations[104] dealing with applications made under the Environmental Protection Act 1990 require that HMIP consults with the NRA in the case of discharges to controlled waters, and the sewerage undertakers in the case of discharges to sewer. This procedure will undoubtedly assist HMIP but it remains possible that the requirements of Article 5 are not being fully complied with.

(5) Such authorizations are to be granted for a limited period only. [Article 3, 76/464/EEC]. The authorizations referred to in Article 3 of 76/464 are to be reviewed at least every four years. [Article 3, 86/208/EEC].

Authorizations granted under the provisions of the Environmental Protection Act 1990 are required to be reviewed every four years[105]. There is no such requirement in the Water Industry Act 1991 nor in the Water Resources Act 1991. It remains with the regulators therefore to see that such reviews take place. If one considers Article 3 of 86/280/EEC (which requires authorizations to be reviewed at least every four years) in the light of Article 3 of 76/464/EEC, to which it refers (which requires that authorizations shall be granted for a limited period only), it may be argued that authorizations should be time limited ie. expire automatically rather than merely require some form of review of the extant authorization.

(6) If the discharger states that he is unable to comply with the required emission standard, or if this situation is evident to the competent authority, authorization is to be refused. [Article 5(3) 76/464/EEC].

This condition is mirrored in Section 6(4) of the Environmental Protection Act 1990 which requires that "An application shall not be granted unless the enforcing authority considers that the applicant will be able to carry on the process so as to comply with the conditions which would be included in the authorization."

Responsibilities placed on the NRA concerning likely non-compliant discharges are set out in a Direction from the Secretary of State[106]. This Direction requires that the NRA refers to the Secretary of State any applications for discharges to controlled waters where the granting of such applications would result in the breach of water quality objectives. There are no specific duties placed on the water companies with regard to likely non-compliant discharges, although they will wish to control trade effluent discharges to sewer to ensure compliance with their own sewage works consent conditions. They will similarly wish to control the release of, for example, heavy metals to sewer in order to protect their sewage sludge disposal routes (this is particularly important where sludge is disposed of to agricultural land). Nevertheless, the water companies are at liberty to impose whatever conditions they wish (subject to appeal) and to enforce, or not, such conditions. There are no specific obligations placed on HMIP in the event of referrals being made to the Secretary of State (under the provisions of the Water Industry Act 1991) of such proposed discharges of List I substances.

(7) Action is to be taken by the competent authority in the event of failure to comply with the conditions of authorization. [Article 5(4) 76/464/EEC].

The regulatory arrangement for the control of discharges of List I substances to sewer is complex.

Conditions imposed by HMIP, on behalf of the Secretary of State, are incorporated into a trade effluent consent issued by the sewerage undertaker. Thus, both HMIP and the sewerage undertaker have powers to enforce limits set on List I substances. The sewerage undertakers are under no obligation to enforce conditions contained within their consents and will only tend to do so where non-compliance is causing them some operational problems, e.g. non-compliance with their own sewage works consent, damage to the sewers, or treatment difficulties. Discharges to controlled waters that are consented and controlled by the NRA are the subject of written, internally agreed, enforcement procedures. This involves the taking of formal samples (evidence from which can be used in court), the sending of written warnings, and finally prosecution. HMIP has no stated enforcement policy with regard to List I substances.

(8) Limit values are to apply except in those cases where a Member State can prove to the Commission, in accordance with monitoring procedures set up by the Council, that quality objectives are being met. [Article 6, 76/464/EEC].

This provision has not been implemented into all of the appropriate controlling UK law, and this has significant consequences, as discussed below in the "Hays Chemicals" case study.

(9) Programmes are to be established and implemented by Member States to reduce pollution by List II substances (which includes all List I substances for which no specific standards have been established). [Article 7, 76/464/EEC].

The Directive is not specific in its requirements as to the precise form that any such programme should take. The United Kingdom implements its obligations, in so far as they relate to List II substances, through a national programme for pollution reduction which applies generally to such substances, rather than by separate programmes for individual substances.

The essential features of this reduction programme are:

(1) a system of prior authorization of discharges via the consenting requirements of the Water Resources Act 1991;

(2) the determination of, and adherence to, nationally derived water quality standards;

(3) monitoring and measurement of these substances and assessment of reductions in riverine loads as a result of these controls;

(4) other measures in respect of individual substances and diffuse sources.

Under the provisions of the Water Resources Act 1991, all direct discharges (except for those that arise from prescribed processes, authorized under the provisions of the Environmental Protection Act 1990) require prior consent from the National Rivers Authority. Conditions attached to such consents are calculated in order to ensure that the relevant downstream environmental quality objectives are met, having regard to the use(s) of the receiving water.

Under the provisions of the Water Industry Act 1991, consent limits, set on discharges to sewer of List II substances, are determined having regard to:

(1) the potential for damage to the fabric of the sewer;

(2) the local sewage sludge disposal strategy (acceptable levels of heavy metals in sewage sludge for disposal to agricultural land are laid down in European legislation);

(3) the potential for inhibition of the sewage treatment process at the receiving sewage treatment works;

(4) the need for compliance with the water company's own NRA consent to discharge from the sewage treatment works.

Under the Environmental Protection Act 1990, discharges of all substances from prescribed processes are controlled by virtue of an authorization issued by HMIP. One of the objectives of control of such processes is that releases from the process of scheduled substances (for water these are essentially List I substances) are minimised and rendered harmless, and that other releases are rendered harmless, using best available techniques not entailing excessive cost (BATNEEC). Control under the Environmental Protection Act 1990 is seen as fundamental to the United Kingdom's reduction programme required under this Article.

The List II reduction programme also includes the development of national environmental quality standards. A number of such standards have already been developed for, for example, the List II metals (limits, in some cases, having been derived from the directive on surface water intended for use as drinking water[107] and the directive on the quality of freshwaters needing protection or improvement in order to support fish life[108]). These standards are currently promulgated via Department of the Environment Circulars and have, therefore, no statutory basis, although copper and zinc are now the subject of control under the recent surface water classification regulations[109].

(10) Member States are to produce an inventory of discharges containing List I substances. [Article 11, 76/464/EEC].

Each year the Department sends to the Commission an inventory of discharges that are authorized for List I substances. This list is initially compiled by the

NRA using information from their own consents register and information as supplied by the sewerage undertakers under the Sewerage Undertakers (Information) Direction 1991. Such information includes:

(a) the date of the consent or agreement;

(b) the name of the occupier of the premises to which the consent or agreement relates;

(c) a brief description of the process to which the consent or agreement relates and the production capacity of that process, expressed in tonnes per year;

(d) the name of the waters into which effluent to which the consent or agreement relate is ultimately discharged;

(e) the Ordnance Survey Grid reference of the place at which such effluent is so discharged;

(f) if such effluent is received by a sewage treatment works, the name of that works; and

(g) particulars of at least two of the following –
  (i) the rate of discharge permitted under the consent or agreement, expressed in cubic metres per day (period of 24 hours);
  (ii) the maximum concentration expressed as a monthly or daily average in microgrammes per litre;
  (iii) the authorised [sic] load expressed in kilogrammes per month."[110]

The NRA needs such information in order to consider what limits are required in consents granted for the discharges from the sewage works receiving these effluents. However, it is questionable whether such reporting should be via the NRA or whether a more sensible approach would be for the sewerage undertaker to report via the "competent" authority, HMIP.

It should be noted that the reporting requirement placed on the sewerage undertaker only relates to consented discharges. Unconsented discharges would remain unreported.

As processes that are prescribed for control under the Environmental Protection Act 1990 become authorized, the NRA will revoke its own consents (provided that all the discharge from controlled waters arises from the authorized process). As such these consents will no longer appear on the NRA List I inventory. HMIP has not been issued with any formal instruction requiring it to furnish information on authorized releases of List I substances to the NRA or the Department of the Environment. It is possible that the NRA will extract the necessary information from its public register (which contains details of authorizations granted by HMIP), but unless this administrative requirement is adopted in some consistent and formal manner the inventory will in time become progressively incomplete.

(11) Member States are to supply the Commission with information concerning

(a) details of authorizations granted

(b) details of the inventory

(c) results of monitoring [Article 13, 76/464/EEC].

Parts (a) and (b) have been discussed above.

The situation concerning monitoring remains unchanged, even as processes pass from NRA control under the Water Resources Act 1991 to HMIP control under the Environmental Protection Act 1990. Compliance with environmental quality objectives in controlled waters will still be demonstrated by virtue of sampling programmes carried out by the NRA. This compilation of data is reported each year to the Commission by the Department of the Environment.

(12) New and substantially increased discharges are only to be authorized if the best technical means are used to eliminate and minimise the release of the substance. [Article 3(3) 82/176/EEC]

An important departure from the Environmental Quality Objectives approach occurs in the case of new or substantially changed plant. A dispute over this arose during discussions leading to Directive 82/176/EEC on mercury from the chlor-alkali industry[111]. The French and Italian governments argued that if a country applying environmental quality objectives did not have to apply limit values that would otherwise have applied to new plants (under the Uniform Emission Standards approach), there would be distortion of competition.

The Commission resolved the problem by including wording in the Directive requiring that new and substantially changed plant should only be authorized if it used the "best technical means available" for minimising releases of List I substances. This requirement places an obligation on the regulators in England and Wales to:

(1) identify new plant; and

(2) identify substantial increases in the handling capacity of existing plant.

Many of the sources of List I substances are now prescribed for control under the Environmental Protection Act 1990, and as such require authorization to operate from HMIP. Non-prescribed processes require consent from the NRA for discharges to controlled waters, and discharges to sewer require the consent of the private water companies (subject to referral to the Secretary of State as previously described). The major sources of List I substances will, therefore, be adequately controlled, although, as discussed above, there are some concerns about the mechanism for identifying new sources.

However, the situation regarding increases of existing discharges is more problematic. Department of the Environment Circulars 18/85 and 7/89 suggest that a practical interpretation of "substantially changed" should be taken as meaning a 20% increase in handling capacity of the relevant substance. It was common practice in the days when the regional water authorities regulated such discharges for the local inspectors to exchange correspondence with the discharger effectively agreeing a handling capacity. A "base" was agreed, therefore, against which to assess future increases in handling capacity, the base year being taken as the year of implementation of the directive concerned. The Sewerage Undertakers (Information) Direction 1991 requires that the water companies supply certain information each year to the Secretary of State. This Direction requires that information is provided on "production capacity", that is to say, the actual capacity of the plant to produce. This is clearly different from handling capacity, which by contrast relates to the actual throughput of the plant. For example, a site could have an installed production capacity well in excess of its current handling capacity. This would allow for a substantial increase in the throughput of dangerous substances of the plant without making any change in its "production capacity".

The practice of agreeing handling capacity is not carried out within HMIP nor does HMIP have access to the previously agreed handling capacities. Given the information that is supplied to the Secretary of State, it is unclear how HMIP would determine whether a site has substantially increased its handling capacity.

Under the Environmental Protection Act 1990 procedures for authorization one would suppose that an applicant would provide details of handling capacity as part of his application (although this is not a formal requirement). One would also assume that as standards are tightened to those contained in the Chief Inspector's Guidance notes (standards which reflect the best available techniques for elimination or minimisation of releases), processes will all ultimately be applying "best technical means available". The sources which are not adequately covered, in terms of the Directive requirements, are those sources that discharge to sewer but do not fall for regulation under the Environmental Protection Act 1990. In these cases, unless a substantial increase in handling capacity resulted in the need for a review of the extant trade effluent consent (giving rise to a referral to the Secretary of State under the provisions of the Water Industry Act 1991), the requirement to use the "best technical means available" may not be met. The NRA is now proposing to include a condition in sewage treatment works consents requiring that the SU notifies it of the receipt of any new or increased release of List I substances into the sewerage system[112]. Such a condition may help the situation but, until some formal system of exchange of information between the NRA and HMIP is set in place, the United Kingdom government remains at risk of breach of certain requirements of this Directive.

It should also be noted that information required under the provisions of the Sewerage Undertakers (Information) Direction 1991 specifically relates to substances prescribed by The Trade Effluents (Prescribed Processes and Substances) Regulations 1989, as amended by The Trade Effluents (Prescribed Processes and Substances) (Amendment) Regulations 1990. The substances prescribed under these regulations include two of the four recently notified List I substances, 1,2 dichloroethane and trichlorobenzene, but not perchloroethylene or trichloroethylene (which are prescribed under The Trade Effluents (Prescribed Processes and Substances) Regulations 1992). This means that the sewerage undertakers are not required to furnish the Department of the Environment with any information relating to two List I chlorinated solvents.

It has been shown, therefore, that whilst some fundamental obligations of the Dangerous Substances Directives have been adequately implemented, others have to a greater or lesser extent not been adequately implemented. The question arises whether the inadequately implemented obligations are covered by the concept of direct effect.

It is the author's contention that the requirements for direct effect ie. that obligations placed on the Member States in Directive 76/464/EEC, are "clear, precise and unconditional", are indeed met. Kramer concurs with this view and states, "A first group of provisions with direct effect in the light of Court rulings concerns provisions that lay down maximum values for permissible discharges, ie. limit values, emission values etc . . . These values are precise, unconditional and not dependent on further action by the Council or the Member States. Consequently, they have been applicable since the beginning of 1989 in the Community even where a Member State has not taken corresponding implementing measures."[113]

9. CASE STUDIES AND INDUSTRY'S VIEW OF THE IMPLEMENTATION OF 76/464/EEC

1. ENVIRONMENTAL QUALITY OBJECTIVES AND EMISSION STANDARDS: "THE CASE OF THE CHLOR-ALKALI INDUSTRY"

In 1983 the Department of the Environment sent to the Commission the first annual return required by the directive 82/176/EEC (on limit values and quality objectives for mercury discharges by the chlor-alkali electrolysis industry), listing the five operational plants in Britain and setting out details of their consents. The five plants were:

(1) Castner-Kellner Works, Runcorn (ICI) discharging into the Weston canal and affecting the Mersey Estuary and Liverpool Bay;

(2) Hillhouse Works (ICI) discharging into the Wyre Estuary and affecting Morecambe Bay and Liverpool Bay;

(3) Ellesmere Port (Associated Octel) discharging into the Manchester Ship Canal and affecting the Mersey Estuary and Liverpool Bay;

(4) Sandbach Works (Hays Chemicals) discharging into the Trent and Mersey Canal affecting the Mersey Estuary;

(5) Staveley Works Chesterfield (Staveley Chemicals) discharging into the River Rother and affecting the River Don[114].

In the case of the first three plants, limits were to be set such that environmental quality objectives would be met. Improvements to discharges from the Staveley Works were to result in the attainment of the environmental quality objective in the receiving river by 1986. The problems at the Sandbach works were not, however, so easily resolved and these are discussed more fully below.

2. "THE CASE OF HAYS CHEMICALS"

One of the essential features of the Council Directive 76/464/EEC on pollution caused by the discharge of dangerous substances is embodied in Article 6. This Article states that "The limit value established in accordance with paragraph 1" (of Article 6) "*shall* (author's emphasis) apply except where a Member State can prove to the Commission, in accordance with a monitoring procedure . . . that the quality objectives . . . are being met and continuously maintained." Kramer concurs and states that "Article 6(3) of Directive 76/464 permits Member States in exceptional cases to establish quality objectives in place of limit values."[115]

Such a system of control was recognised within England and Wales during the regulatory tenure of the water authorities, and, more recently, during the tenure of the NRA. Paragraph 9 of the Department of the Environment Circular 18/85 states "If, exceptionally, there are cases where the appropriate quality standards* cannot be initially met, water authorities may wish to use the limit value approach until such time as the quality objectives can be achieved. In any such cases the Department of the Environment's agreement must be sought." A Department of the Environment consultation paper on the revision of Circular 18/85 states that it is "the Department of the Environment's and the Welsh Office's intention that, in the UK, the EQO approach should *wherever possible* (author's emphasis) be used for List I as well as List II substances."

Following a Reasoned Opinion from the Commission in 1988, this Circular was superseded by Circular 7/89, which updated and added to the previous advice. Haigh explains that "The Commission had cited the UK government's failure in the case of the daughter directives on HCH and cadmium adequately to specify quality objectives, sufficiently clear authorization procedures and the required methods of analysis."[117]

Paragraph 14 of the subsequent Department of the Environment Circular 7/89 still maintained "If exceptionally there are cases where the appropriate quality objective cannot be initially met, competent authorities may need to apply the limit value approach until such a time as the quality objectives can be achieved. In any such cases, the department should be informed at an early stage."

This system of control outlined in these circulars, although reinforcing the UK Government's commitment to environmental quality objectives, mirrors the dual control approach required by the Directive.

In 1989 (and as amended in 1992) a system for the classification of surface waters was set up which laid down a set of criteria established on the basis of the requirements of the Dangerous Substances Directive and its daughter directives[118]. The criteria set down in this system were established as statutory water quality objectives by virtue of a Notice issued pursuant to section 105(1) of the Water Act 1989.

The Direction issued to the NRA, pursuant to sections 146 and 149 of the Water Act 1989, sets out the duty of the Authority. As well as ensuring that statutory water quality objectives are met, the Direction also states that "Where the Authority is of the opinion, as respects any waters in relation to which it has notified the Secretary of State . . . that the steps so mentioned are unlikely to secure the restoration within 12 months from the date of the notification, of the quality of the waters to that required for the relevant classification, it shall provide the Secretary of State with such information as will enable him, where appropriate, to determine in relation to any relevant discharge to those waters an appropriate emission standard in accordance with the relevant Council Directive."

It can be seen, therefore, that the Department of the Environment still recognises the applicability of the use (albeit on a temporary basis) of the Uniform Emission Standards approach. It is the author's contention that this is a requirement of the framework Directive 76/464/EEC.

This dual approach, permitted under previous legislation, is no longer allowed under the provisions of the Environmental Protection Act 1990. Under section

* The terms environmental quality objective and environmental quality standard are sometimes used interchangeably and EC and UK terminology differ. In the UK an environmental quality objective defines the use or uses, as identified by the competent authority, for which a body of water must be suitable; the environmental quality standard is the concentration of a substance in the receiving water which must not be exceeded if the environmental quality objective is to be realised.[116]

28(3) of the 1990 Act, HMIP is not permitted to authorize a process if the NRA certifies that it "will result in or contribute to" a breach of a statutory water quality objective. It should be noted that this essentially removes the option for the adoption of the Uniform Emission Standards approach for those processes – and only those processes – which require authorization under Part I Section 2 of the 1990 Act. All processes that remain outside such control will enjoy the benefit of control under the Water Resources Act 1991, which, as has been explained above, more accurately mirrors the system of control set out in 76/464/EEC. One company has suffered as a result of this regulatory arrangement.

Hays Chemicals Distribution operates five processes at their Sandbach site, one of which entails the manufacture of chlorine using mercury cells. The plant is highly efficient in terms of mercury lost per tonne of chlorine produced, easily meeting the emission standard set out in Directive 82/176/EEC[119] for such plant, and the mercury recovery plant produces an effluent with a mercury concentration of the order of that which is acceptable in drinking water. The combined effluent from the site, which includes contaminated surface water from historic operation, discharges into the Trent and Mersey Canal which affords a limited dilution of about 15 to 1.

In the early 1980's, and in accordance with provisions laid down in Circular 18/85, North West Water Authority informed the Department of the Environment of the fact that the Trent and Mersey Canal downstream of Hays Chemicals was failing to comply with the environmental quality objective as laid out in 82/176/EEC (on limit values and environmental quality objectives for discharges by the chlor-alkali electrolysis industry). Instead of being required to comply with the environmental quality objective, the site was allowed to continue to operate under the system of emission standards as allowed for in the Directive.

On 2 January 1990 the Secretary of State for the Environment notified the NRA, under section 105 of the Water Act 1989, of the establishment of water quality objectives for all inland waters. The quality objective so notified was the satisfaction by those waters on and after January 1990 of the requirements for classification "DS1". That classification was established by the Surface Waters (Dangerous Substances) (Classification) Regulations 1989 (SI1989 No 2286). The concentration of mercury and its compounds in the relevant inland water necessary to satisfy the classification DS1 is set out in those regulations as a maximum annual mean concentration of 1 μg/l. The making of those regulations and the notification of the water quality objectives were required in order for the UK government to comply with its obligations under 82/176/EEC.

The Regulations themselves do not mention the Uniform Emission Standards approach. The Direction issued under Section 146 of the Water Act 1989, which sets out the duties of the NRA in so far as they relate to the Regulations, does however recognise this approach. This Direction requires the NRA to inform the Secretary of State of those circumstances where water quality objectives laid down in the Regulations will not be met. Under these circumstances the NRA is required to provide the Secretary of State with information to enable him to determine an appropriate emission standard in accordance with the relevant Council Directive.

The current discharges from the Hays chemical plant of between 15–20 μg/l of mercury may, taken alone, result in compliance with the environmental quality objective. Unfortunately, and possibly due in part to re-suspension of historic bed deposits by periodic dredging operations, the environmental quality objective is breached. The process operated at this site is prescribed for control by HMIP[120] and as such, if Hays Chemicals wish to continue to operate this process, an authorization from HMIP is required[121]. As part of the application procedure, there is a statutory requirement to consult with the NRA.

Section 28(a) of the Environmental Protection Act 1990 requires that "the enforcing authority shall not grant an authorization under this Part if the National Rivers Authority certifies to the enforcing authority its opinion that the release will result in or contribute to a failure to achieve any water quality objective . . ." The current quality of the water in the Trent and Mersey canal is a matter of public knowledge, with analytical data being available from the NRA public register. Because the outlet from the site has recently been relocated there is currently insufficient monitoring data to ascertain whether the environmental quality objective (which is based on an annual mean concentration) is being breached. As such the NRA will be unable to certify that the discharge is causing or contributing to a failure of the environmental quality objective, and an authorization from HMIP can be granted. The NRA statutory consultee response is reproduced in Figure 6. However, if monitoring currently being carried out by the NRA subsequently demonstrates that the environmental quality objective is being breached, the enforcing authority (HMIP) is required to vary the conditions of the authorization as required[122].

This operator has been directly disadvantaged by changes in legislation that have removed one of the options, the Uniform Emission Standards approach, which is permitted by the Directive.

Article 10 of the Directive allows Member States to apply "more stringent measures" than those provided for in the Directive. The wording of paragraph 3 of Article 6 may be interpreted as imposing a specific obligation on Member States to adopt emission standards where environmental quality objectives are not met. It is difficult to argue the case that adoption of control by environmental quality objectives is a more stringent measure where:

(1) it appears still to be an option for those processes not regulated under EPA90; and
(2) where discharges to large estuarine waters can, as a result of the Environmental Quality Objectives approach, release levels of dangerous substances far in excess of those that would be allowed under the Uniform Emission Standards approach.

The consent limit that has now been imposed on the discharge from Hays Chemicals has been calculated on the basis of current effluent quality. As such this limit is more stringent than that which had previously been imposed by the Uniform Emission Standards approach. The question remains therefore whether Member States can adopt "more stringent measures" on a site by site basis (and presumably thereby place individual companies at a competitive disadvantage) or whether such measures have to be applied nationally.

It would be open to this company to take action against the United Kingdom government for what appears to be an incorrect implementation of directive provisions. Such incorrect implementation could be regarded as having removed the legal certainty offered by the Directive which clearly sets out the regulatory system that is to operate for the control of dangerous substances. Such action would be taken in the national courts and, as indicated previously, would be for damages of a tortious standard.

Date 11th March 1994

NRA

National Rivers Authority
North West Region

HMIP
Stockdale House
8 Victoria Road
Headingley
LEEDS
LS6 1PF

Dear Ms Atkinson

ENVIRONMENTAL PROTECTION ACT 1990, Part I.
APPLICATION FOR AUTHORISATION REFERENCE NOS. AL9335/AL9343/AL9327/AL9319.
OPERATOR: HAYS CHEMICAL DISTRIBUTION LTD.

Thank you for sending me a copy of the above application for authorisation, as required under the consultation procedures described in The Environmental Protection (Applications, Appeals and Registers) Regulations, 1991.

The information you have provided will be held on the Water Resources Act 1991 Public Register, in hard copy format, and made available for public inspection on request during normal working hours.

I would make the following comments:

You will be aware from our earlier discussions that we have been concerned for some time about the levels of mercury present in water samples taken from the Trent and Mersey Canal downstream of the Hays Chemicals discharge. The NRA has a statutory responsibility under section (84) of the Water Resources Act 1991 to ensure that the environmental quality standard for mercury is achieved and maintained in the canal.

Since last December, we have carried out intensive monitoring of the canal to determine the impact of mercury discharges from Hays Chemicals. Data collected so far shows that mercury levels in the canal at Crows Nest Bridge SJ 7305 6211, a point approximately 1000 metres downstream of the current discharge point, are such that the EQS of 1.0 microgram per litre (annual mean concentration) is threatened. However at this stage we do not have a full 12 month data set to enable us to assess compliance with the EQS. Data collected over the past 3 months shows a mean concentration of 0.91 micrograms per litre. I enclose a public register retrieval showing the results of our monitoring.

During the winter months there is reduced boating activity on the canal and consequently less water movement. It may be that mercury levels will fall in the spring and summer months as canal traffic increases but this cannot be guaranteed.

In view of the fact that average mercury levels in the canal are close to the 1.0 microgram per litre EQS, it is important that every effort is made by the Company to reduce their mercury emissions to canal. In the meantime we will continue with our enhanced monitoring programme at Crows Nest until we have sufficient data to determine whether or not the EQS, as calculated on an annual mean basis, is exceeded. If this proves to be the case then we will be required to inform the Department of the Environment, as part of our statutory reporting, and furnish them with reasons for non-compliance, together will details of actions to be taken to secure future compliance. Such actions may require a review of mercury conditions specified in the HMIP authorisations.

I enclose a copy of the extant consents which covers discharges from the site including any effluent stream from the process for which application has been made for authorisation. Any requirements imposed by an authorisation issued by HMIP must ensure that the conditions contained in our consent are complied with.

Also enclosed is a copy of our pollution prevention guidelines. General site housekeeping measures associated with storage, loading, vehicle washing etc., should conform with these guidelines.

Details of our sampling regimes associated with the discharge and environmental monitoring of the receiving water are attached. The Memorandum of Understanding requires that our organisations agree on a sampling programme for the effluent and receiving water.

If you have any comments then please contact either Glyn Edwards or Alison Bradley

Yours sincerely,

Mr C.M. Newton
Environmental Quality Manager
National Rivers Authority North West Region

Figure 6 NRA statutory consultee response

3. "THE CASE OF BELMONT BLEACHING AND DYEING"

In the late 1980's several Lancashire rivers were found to contain pentachlorophenol (PCP) above the permitted environmental quality objective limit[123]. It was found that the source of the problem was due to scouring, by the textile industry, of imported PCP impregnated cotton fabric. The problem was found to be due to the fact that PCP was being added to the starch-based size, in the country of origin of the cotton fabric (India, the Far East, Pakistan, Egypt etc) prior to the weaving of the cotton. PCP was being added to the size to prevent its deterioration in the hot climates of these countries and to prevent mildew on the cloth during its transport by ship[124]. The alkaline scouring processes, which are carried out to clean the cloth prior to dyeing, solubilise the PCP which passes in the trade effluent discharge to the receiving sewage treatment works and on into the local watercourse.

By mid-1991 many companies had taken steps to reduce their discharge of PCP to sewer, but one Bolton firm, Belmont Bleaching and Dyeing, had been less successful in its efforts. Belmont Bleaching and Dyeing is a small textile works located on the headwaters of the Eagley brook in Lancashire. Its liquid textile finishing waste comprises about 80% of the flow into Belmont sewage treatment works, with the residual 20% of the flow arising as domestic waste from the village of Belmont. Textile effluent contaminated with PCP passed through Belmont sewage works and into the Eagley brook, which consequently failed to meet the required PCP environmental quality objective by some considerable margin. The NRA announced its intention to prosecute North West Water plc for its breach of the 6 μg/l PCP limit set in the final effluent consent for the sewage works. The discharge of trade effluent to Belmont sewage works is made by virtue of an agreement, a form of contract made between the discharger and the receiving sewerage undertaker under Section 129 of the Water Industry Act 1991. Unfortunately, due to the nature of the agreement between North West Water and Belmont Bleachers and Dyers, North West Water was powerless to control the PCP content of the influent to its sewage works and thereby had little control over the PCP content of the sewage works effluent.

In June 1991 HMIP varied the agreement between North West Water and Belmont Bleachers and Dyers, inserting a specific provision controlling PCP in the trade effluent. The limit set required Belmont Bleachers and Dyers not to exceed 350 μg/l of PCP (100 μg/l as a daily average), with a tighter limit of 6 μg/l coming into force twelve months later. North West Water argued that the immediate standard would not provide adequate control to ensure compliance with its own 6 μg/l sewage works consent limit and as a consequence applied to the NRA for a variation of the sewage works PCP limit to 100 μg/l. In view of the fact that the discharge from the sewage works was to a watercourse that was failing its environmental quality objective, the NRA referred the matter to the Secretary of State for determination.

As far as can be established from the public register this matter was never determined by the Department of the Environment, and North West Water was successfully prosecuted for its failure to comply with the 6 μg/l consent limit[125]. The interesting point here is that the Department of the Environment indicated that, in the short term, whilst Belmont Bleachers and Dyers were to be permitted to discharge at 100 μg/l, the Uniform Emission Standards approach was to be applied[126].

It could be argued that, in this case, North West Water could have taken steps against the Secretary of State for non-determination of its application to vary its consent. Had an emission standards approach been adopted with regard to the discharge from the sewage works, the prosecution of North West Water would not have succeeded.

4. "THE CASE OF THE YORKSHIRE LEATHER INDUSTRY"

Directive 84/491/EEC, on limit values and quality objectives for discharges of hexachlorocyclohexane, was the first directive for a pesticide that had a polluting effect mainly from diffuse rather than point sources[127]. The Directive sets limit values for production and extraction plants and also requires Member States to fix emission standards for other industrial sectors "treating" HCH[128]. The Environmental Quality Objectives approach, as explained previously, requires all releases of the controlled substances to be regulated. However, the imprecise meaning of "treating" allows Member States following the Uniform Emission Standards approach to exclude some sources of HCH.

The problem has been particularly highlighted in Yorkshire where the Rivers Aire and Calder have been found to contain levels of HCH well in excess of the required environmental quality objective[129]. The source has been traced to the washing of fleeces, and although these releases are passive (as a result of contamination of raw materials), rather than active, the Environmental Quality Objectives approach makes the control of such releases necessary. Elsewhere in the Community such plants are not regulated and this has given rise to complaints by the leather industry of unfair competition.

Again it is possible to argue that the framework directive *requires* the Uniform Emission Standards approach to be applied where environmental quality objectives are not met. As such the leather industry would be able to take action in the national courts against the NRA for incorrect implementation of directive obligations.

5. INDUSTRY'S VIEW OF IMPLEMENTATION OF 76/464/EEC

Part of the research associated with this work was to seek the views of industry on the method of implementation of the Dangerous Substances Directive in England and Wales. These views were sought in a letter which was sent to a number of major trade associations, organizations that represent a cross section of industry affected by the regulation of dangerous substances.

The response to the letter was very disappointing with replies only being received from "The Metal Finishing Association" and the "Textile Finishers Association".

The CIA apologised for not being able to reply due to pressure of other work.

The two responses received both expressed a view that the preferred system of regulation is that of emission standards. The transparency and level of certainty of control given by the Uniform Emission Standards approach was felt to offer advantages over that of control by environmental quality objectives. Both respondents felt that the Environmental Quality Objectives approach was placing their industry at a competitive disadvantage when compared to industry in the rest of Europe.

This is not surprising as the type of industry represented by these two trade associations is typically located inland. By contrast, it is the major chemical works located on estuaries, with resultant large receiving water dilutions, that benefit from the Environmental Quality Objectives approach. The chemical sector of United Kingdom industry has the more significant political voice and was able to influence the government in its choice of a control regime when the draft Dangerous Substances Directive was being negotiated (although the water industry also supported the adoption of environmental quality objectives).

10. CONCLUSIONS AND RECOMMENDATIONS

1. CONCLUSIONS

The doctrine of direct effect is now well established in European law. It means that individuals can rely on the provisions of directives, even where those provisions have not been implemented, or have been implemented incorrectly in national law. All that is required for direct effect is that the provisions in question confer rights which are clear, precise, and unconditional.

Moreover, where individuals suffer damage as a result of incorrectly implemented directive provisions, they are entitled to seek damages against the relevant body.

In the case of regulation of discharges of dangerous substances to the aquatic environment the relevant body would be a private water company, or the NRA or HMIP.

The Dangerous Substances Directive contains provisions that are generally regarded as being clear, precise, and unconditional, but there are questions regarding the implementation of these provisions.

First, the nature of the substance being controlled (particularly concerning isomers) depends upon which piece of implementing legislation is being used. No single control regime, be it the provisions of the Water Industry Act 1991, or the Environmental Protection Act 1990, correctly defines the complete list of substances for which daughter directives have been produced. As such, in this respect, the regulators in England and Wales find themselves imposing faulty legislation and are ignoring the direct effect doctrine.

A second point of concern relates to the method of implementation of European directives. The United Kingdom government has adopted a system of control based on environmental quality objectives. One provision of the Dangerous Substances Directive requires that emission standards are applied where environmental quality objectives are not met. This provision may be regarded by industry as being a "right" which is conferred by the Directive. In order for such a provision to confer rights on individuals it must be considered whether this provision had as its intention the conferral of rights or whether the issue is only that of incorrect implementation of directive provisions.

It may be argued that one right conferred by this provision is that of legal certainty, that is to say that dischargers of substances, subject to the provisions of the Directive, operate within a clearly defined regulatory system as set out in the Directive. This regulatory provision has been denied to certain operators, within England and Wales, as a result of a failure to recognise the option of regulation by the Uniform Emission Standards approach.

The Directive allows Member States to impose tighter standards than those required by the Directive and its daughters. However, the use of this argument to support the removal of the option to regulate by emission standards cannot be supported for the following reasons:

(1) standards required by the Environmental Quality Objectives approach are frequently more relaxed than those that would be required under the Uniform Emission Standards approach, particularly for discharges to estuaries and coastal waters;

(2) legislation still allows for industry that does not fall for regulation under the Environmental Protection Act 1990 to be controlled using the Uniform Emission Standards approach;

(3) it is only the wording of the Environmental Protection Act 1990 that precludes the Uniform Emission Standards approach.

A consequence of this is that the regulators are liable to compensation claims from any party which suffers damage from their actions in this area.

The Environmental Quality Objectives approach has been applied to individual dischargers who previously operated under the Uniform Emission Standards approach. It may be argued by the regulators that change in regulatory control constitutes a "more stringent measure" and is, therefore, allowed under Article 10 of 76/464/EEC. The question remains whether such an approach can be implemented on a site-by-site basis or whether such measures are to be adopted by the State in its broadest sense, that is to say by way of more stringent national measures.

Finally, there are areas which are inadequately covered by formal procedures. For example, the provision of information to the Commission is in danger of being undermined as a result of processes passing out of NRA control into HMIP control. There is no formal administrative requirement on HMIP to furnish the Department of the Environment with necessary information on authorization of releases of List I substances. It is hoped that this issue of transfer of information between the different regulatory bodies in England and Wales will become more formalised under the proposed Environment Agency.

From a practical view point it would seem unlikely (although not impossible) that individuals would pursue actions against the United Kingdom government for incorrect implementation of the Dangerous Substances Directive and its daughters. However, pressure groups such as angling associations, environmental groups, and industrial trade associations, certainly have the financial resources to pursue any perceived deficiency in implementation via the national courts. As the doctrine of direct effect becomes increasingly recognised as a viable basis for litigation in national courts one could expect such groups to become more active in this area. However, as the National Smokeless Fuels Ltd Monkton Works case demonstrates, where legislation is in place to implement directive provisions, reliance on direct effect is unlikely to be received sympathetically, at least by Her Majesty's Ministers of State.

2. RECOMMENDATIONS

- There is an immediate need for regulators of discharges of dangerous substances to the aquatic environment to be made aware of the doctrine of direct effect.
- As a minimum, regulators should implement those requirements that, although not specifically mentioned in extant legislation, are nevertheless required by the Dangerous Substances Directive and its daughters. They should also follow the approach being adopted by the national courts and interpret legislation in the light of directive requirements.

- Regulators must press urgently for changes to the law where extant legislation is perceived as being contrary to directive requirements.
- Legislators must adhere more closely to the wording of directives when drafting implementing legislation.
- The Department of the Environment should consider imposing a formal requirement on HMIP to furnish, either directly or via the NRA, that information which is required for the compilation of the United Kingdom's return to the Commission.

The issue of identification of new releases will increase in importance as future daughter directives are issued and implemented. The system for identification of such releases, and the notification of details of such discharges to the Commission, needs to be urgently examined. The formation of the Environment Agency must be seen as an opportunity to review the United Kingdom's approach to compliance with the dangerous substances directives.

References

1. C250/39 Official Journal of the European Communities C250 Vol. 35 28 September 1992.
2. Environmental Resources Management Limited, Contract Reference B4-3040/015406-92. Administrative Structures and Implementation of the Community Directives on the dangerous substances discharged into the aquatic environment.
3. M Sheriden, unpublished Notes. UMIST.
4. M Sheriden, unpublished Notes. UMIST.
5. M Sheriden, unpublished Notes. UMIST.
6. Council Directive 76/464/EEC of 4 May 1976 on pollution caused by certain dangerous substances discharged into the aquatic environment of the Community.
7. Simon Ball and Stuart Bell. Environmental Law p. 52. Blackstone Press Ltd 1991.
8. Simon Ball and Stuart Bell. Environmental Law p. 41. Blackstone Press Ltd 1991.
9. Titanium Dioxide Case: 300/89, Commission v Council (European Parliament supporting the Commission) [Decision 11 June 1991].
10. Simon Ball and Stuart Bell. Environmental Law p. 42. Blackstone Press Ltd 1991.
11. Pascal Fontaine. Europe in Ten Lessons. Luxembourg: Office for Official Publications of the European Communities, 1992.
12. The ENDS Report No. 226 November 1993 p. 39. Published by Environmental Data Services Ltd, Finsbury Business Centre, 40 Bowling Green Lane London EC1R 0NE (Tel: 071-278 4745/Fax: 071-415 0106).
13. Lawrence Collins. European Community Law in the United Kingdom p. 3. Butterworths 1990.
14. Reparation for Injuries case [1949] ICJ Report 174.
15. J H Rayner (Mincing Lane) Ltd v Department of Trade and Industry [1989] 3 WLR 969, [1989] 3 All ER 523, HL.
16. Lawrence Collins. European Community Law in the United Kingdom p. 84. Butterworths 1990.
17. Commission v Italy 91/79 [1980] ECR 1099 and 92/79 ECR 1115.
18. Simon Ball and Stuart Bell. Environmental Law p. 43. Blackstone Press Ltd 1991.
19. Council Regulation (EEC) No. 880/92 of March 1992 on a Community eco-label award scheme.
20. Simon Ball and Stuart Bell. Environmental Law p. 45. Blackstone Press Ltd 1991.
21. Article 3 Treaty on Political Union. Final draft by the Dutch Presidency as modified by the Maastricht Summit.
22. Council Directive 91/173/EEC on restrictions on marketing and use of pentachlorophe No. 1.
23. Council Directive 89/458/EEC on vehicle emissions (small cars).
24. The ENDS Report No. 203 December 1991 p. 22. Published by Environmental Data Services Ltd, Finsbury Business Centre, 40 Bowling Green Lane London EC1R 0NE (Tel: 071-278 4745/Fax: 071-415 0106).
25. A James. "A practical Guide to the Development of Community Legislation". unpublished.
26. EC Commission v Netherlands: 339/87 [1993] 2 CMLR 360.
27. C.E Miller. "Development control as an instrument of environmental management". p. 235 TPR, 61(3) 1990.
28. Circular 7/89. Circular from the Department of the Environment and the Welsh Office. "Water and the environment. The implementation of European Community directives on pollution caused by certain dangerous substances discharged into the aquatic environment".

29. Nigel Haigh. Manual of Environmental Policy p. 4.3-3. Longman.
30. Statutory Instrument 1989 No. 1156. The Trade Effluents (Prescribed Processes and Substances) Regulations 1989. Statutory Instrument 1990 No. 1629. The Trade Effluents (Prescribed Processes and Substances) (Amendment) Regulations 1990. Statutory Instrument 1992 No. 339. The Trade Effluents (Prescribed Processes and Substances) Regulations 1992.
31. The Water Act 1989. Sections 74 and 185(2).
32. The Water Industry Act 1991. Section 138(1).
33. Statutory Instrument 1992 No. 337. The Surface Water (Dangerous Substances) (Classification) Regulations 1992. Notice issued under Section 105(1). Water Quality Objectives for Waters Classified in Accordance with The Surface Water (Dangerous Substances) (Classification) Regulations 1989. Direction to the National Rivers Authority under Section of the Water Act 1989 and relating to Council Directive on Discharges of Dangerous Substances. The Sewerage Undertakers (Information) Direction 1991.
34. The Water Act 1989. Section 105(1).
35. The Water Resources Act 1991. Section 83(1).
36. Statutory Instrument 1994 No. 1057. The Surface Waters (River Ecosystem) (Classification) Regulations 1994.
37. Environmental Protection Act 1990. Section 2.
38. Statutory Instrument 1991 No. 472. The Environmental Protection (Prescribed Processes and Substances) Regulations 1991. Statutory Instrument 1992 No. 614. The Environmental Protection (Prescribed Processes and Substances)(Amendment) Regulations 1992. Statutory Instrument 1993 No. 1749. The Environmental Protection (Prescribed Processes and Substances) (Amendment) Regulations 1993. Statutory Instrument 1993 No. 2405. The Environmental Protection (Prescribed Processes and Substances) (Amendment) (No. 2) Regulations 1993. Statutory Instrument 1994 No. 1271.The Environmental Protection (Prescribed Processes and Substances Etc.) (Amendment) Regulations 1994. Statutory Instrument 1994 No. 1329. The Environmental Protection (Prescribed Processes and Substances Etc.) (Amendment) (No. 2) Regulations 1994.
39. D.M.M Scott. English Law p. 42. Butterworths 1976.
40. D.M.M Scott. English Law p. 62. Butterworths 1976.
41. Foglia v Novello: 104/79 [1980] ECR 745, [1981] 1 CMLR 45.
42. Cilfit Srl v Ministry of Health: 283/81 [1982] ECR 3415, [1983] 1 CMLR 472.
43. Liz Ord. Unpublished paper.
44. Minister of the Interior v Daniel Cohn-Bendit [1980] 1 CMLR 543.
45. Van Duyn v Home Office [1974] 3 All ER 178, [1974] 1 WLR 1107.
46. Lawrence Collins. European Community Law in the United Kingdom p. 76. Butterworths 1990.
47. Article 5 of the Treaty of Rome.
48. Variola v Italian Ministry of Finance: 34/73 [1973] ECR 981.
49. Lawrence Collins. European Community Law in the United Kingdom p. 75. Butterworths 1990.
50. Grad v Finanzamt Traunstein [1970] ECR 825, [1971] CMLR1.
51. Van Gend en Loos v Nederlandse Administratie der Belastingen: 26/62 [1963] ECR 1, [1963] CMLR 105.
52. Lawrence Collins. European Community Law in the United Kingdom p. 16. Butterworths 1990.
53. Pubblico Ministero v Ratti: 148/78 [1979] ECR 1629 [1980] 1 CMLR 96.
54. Francovich v Italian Republic: C-6/90 and C-9/90 [1991] ECR.
55. Josephine Steiner. "From direct effects to Francovich: shifting means of enforcement of Community Law" p. 9. European Law Review Vol. 18 No. 1 February 1993.
56. Josephine Steiner. "From direct effects to Francovich: shifting means of enforcement of Community Law" p. 10. European Law Review Vol. 18 No. 1 February 1993.
57. Bourgoin SA v Ministry of Agriculture, Fisheries and Food [1986] QB 716, [1985] 3 All ER 585, CA.
58. Factortame Ltd v Secretary of State for Transport [1989] 2 WLR 997, [1989] 2 All ER 692, HL.
59. Factortame Ltd v Secretary of State for Transport: C-213/89.
60. Twyford Parish Council and others v Secretary of State for the Environment and aNo.ther [high Court Queen's Bench Division (McCullough J) 26 October 1990, Common Market Law Reports [1992] CMLR 276].
61. Amended proposal for a Council Directive on civil liability for damage caused by waste (91/C 192/04) COM(91) 219 final-SYN 217.
62. Jan H Jans. "Legal Protection in European Environmental Law". European Environmental Law Review June 1993 Vol.2 No. 6 p. 157.
63. Verholen v Sociale Verzekeringsbank: C-87/90 and C-89/90, Rec 1991, I-3757.
64. Commission v Germany: C-361/88 [1991], ECR I-2567.
65. Jan H Jans. "Legal Protection in European Environmental Law". European Environmental Law Review June 1993 Vol. 2 No. 6 p. 158.
66. Marshall v Southampton Area Health Authority: 152/84 [1986] QB 401, [1986] 2 All ER 584, [1986] ECR 723.
67. Lawrence Collins. European Community Law in the United Kingdom p. 89. Butterworths 1990.

68. Johnson v Chief Constable of the R.U.C: 222/84 [1987], QB 129, [1986] 3 All ER 135, [1986] ECR 1651, [1986] 3 CMLR 240.
69. Foster v British Gas plc [1988], 2 CMLR 697, LA.
70. Doughty v Rolls Royce plc [1992] 1 CMLR 1045.
71. Marleasing S.A v La Comercial Internacional de Alimentacions S.A: 106/89.
72. Lawrence Collins. European Community Law in the United Kingdom p. 141–142. Butterworths 1990.
73. A.C Geddes. "Environmental Directives and direct effect in UK national law". The Law Society's Gazette No. 6 14 February 1990.
74. Warner J in National Smokeless Fuels Ltd v CIR [1986] 3 CMLR 227.
75. Rothschild Holdings plc v CIR [1989] 2 CMLR 621.
76. Pickstone v Freeman plc [1988] 3 CMLR 221.
77. Lister v Forth Dry Dock & Engineering Co Ltd [1989] 1 All ER 1134, [1989] 2 WLR 634, HL.
78. Lister v Forth Dry Dock & Engineering Co Ltd [1989] 1 All ER 1134, [1989] 2 WLR 634, HL p. 634.
79. Duke v GEC Reliance [1988] 1 CMLR 719.
80. Webb v EMO Air Cargo (UK) Ltd [1992] CMLR 793.
81. Dekker v Vjv-Centrum: 177/88, [1991] IRLR 27.
82. Shiranikha Herbert. Guardian Law Report 30 July 1994.
83. Times Law Report, 7 January 1994.
84. Richard Macrory. The ENDS Report No. 228 January 1994 p. 44–45. Published by Environmental Data Services Ltd, Finsbury Business Centre, 40 Bowling Green Lane London EC1R 0NE (Tel: 071-278 4745/ Fax: 071-415 0106).
85. C.E Miller. "Coke, Smoke and National Sovereignty: A Case Study of the Role of Planning in Controlling Pollution". Journal of Environmental Planning and Management, Vol. 36, No. 2, 1993.
86. Council Directive 84/360/EEC on the combating of air pollution from industrial plants.
87. C.E Miller. "Coke, Smoke and National Sovereignty: A Case Study of the Role of Planning in Controlling Pollution". Journal of Environmental Planning and Management, Vol. 36, No. 2, 1993, p. 163.
88. Statutory Instrument 1989 No. 1156. The Trade Effluents (Prescribed Processes and Substances) Regulations 1989. Statutory Instrument 1990 No. 1629. The Trade Effluents (Prescribed Processes and Substances)(Amendment) Regulations 1990. Statutory Instrument 1992 No. 339. The Trade Effluents (Prescribed Processes and Substances) Regulations 1992.
89. Statutory Instrument 1989 No. 1156. The Trade Effluents (Prescribed Processes and Substances) Regulations 1989.
90. Water Industry Act 1991. Section 138(2)
91. Water Facts 1993 p. 7–8. Water Services Association Publication.
92. Griffin v South West Water Services Limited. Unreported 1994.
93. S Tromans. The Environmental Protection Act 1990. Text and Commentary p. 43.22. Sweet and Maxwell 1991.
94. COM(92) Vol. II, 27 March 1992, p. 49.
95. William Howarth. "New Strategies for Water Directives" European Law Review October 1992 Vol. 1 No. 4 p. 117–121.
96. Article 3(3) Council Directive 82/176/EEC of 22 March 1982 on limit values and quality objectives for mercury discharges by the chlor-alkali electrolysis industry.
97. Council Directive 80/68/EEC of 17 December 1979 on the protection of groundwater against pollution caused by certain dangerous substances.
98. Water Resources Act 1991. Section 104.
99. Personal correspondence North West Water Authority. Unpublished.
100. Department of the Environment Circular 18/85. The Implementation of Directive 76/464/EEC on pollution caused by certain dangerous substances discharged into the aquatic environment of the Community.
101. Council Directive 90/415/EEC of 27 July 1990 amending Annex II to Directive 86/280/EEC on limit values and quality objectives for discharges of certain dangerous substances included in list I of the Annex to Directive 76/464/EEC.
102. DOE/Welsh Office letter to Chief Executives of Water Authorities in England and Wales dated 25 November 1987 DOE reference: WS/932/65, Welsh Office reference: WEP 12/123/1.
103. Water Industry Act 1991. Section 138(2).
104. Statutory Instrument 1991 No. 507. The Environmental Protection (Applications, Appeals and Registers) Regulations 1991.
105. The Environmental Protection Act 1990. Section 6(6).
106. Direction to the National Rivers Authority under section 146 of the Water Act 1989 and relating to Council Directive on discharges of dangerous substances. January 1990.
107. Council Directive 75/440/EEC of 18 June 1975 concerning the quality required for surface water intended for the abstraction of drinking water in the Member States.

108. Council Directive 78/659/EEC on the quality of fresh waters needing protection or improvement in order to support fish life.
109. Statutory Instrument 1994 No. 1057. The Surface Waters (River Ecosystem) (Classification) Regulations 1994.
110. The Sewerage Undertakers (Information) Direction 1991.
111. Nigel Haigh. Manual of Environmental Policy p. 4.8–10. Longman.
112. Personal correspondence. National Rivers Authority 1994.
113. Ludwig Kramer. "The implementation of Community Environmental Directives within Member States: Some implications of the Direct Effect Doctrine". Journal of Environmental Law Vol. 3 No. 1. p. 42
114. Nigel Haigh. Manual of Environmental Policy p. 4.8–15. Longman.
115. Ludwig Kramer. "The implementation of Community Environmental Directives within Member States: Some implications of the Direct Effect Doctrine". Journal of Environmental Law Vol. 3 No. 1. p. 45
116. Department of the Environment Circular 18/85. The Implementation of Directive 76/464/EEC on pollution caused by certain dangerous substances discharged into the aquatic environment of the Community.
117. Nigel Haigh. Manual of Environmental Policy p. 4.8–12. Longman.
118. (Statutory Instrument 1989 No. 2286 The Surface Waters (Dangerous Substances) (Classification) Regulations 1989). Statutory Instrument 1992 No. 337 The Surface Waters(Dangerous Substances) (Classification) Regulations 1992).
119. Council Directive 82/176/EEC on limit values and quality objectives for mercury discharges by the chlor alkali electrolysis industry.
120. Statutory Instrument 1991 No. 472. The Environmental Protection (Prescribed Processes and Substances) Regulations 1991. Statutory Instrument 1992 No. 614. The Environmental Protection (Prescribed Processes and Substances) (Amendment) Regulations 1992. Statutory Instrument 1993 No. 1749. The Environmental Protection (Prescribed Processes and Substances) (Amendment) Regulations 1993. Statutory Instrument 1993 No. 2405. The Environmental Protection (Prescribed Processes and Substances) (Amendment) (No. 2) Regulations 1993. Statutory Instrument 1994 No. 1271.The Environmental Protection (Prescribed Processes and Substances Etc.) (Amendment) Regulations 1994. Statutory Instrument 1994 No. 1329. The Environmental Protection (Prescribed Processes and Substances Etc.) (Amendment) (No. 2) Regulations 1994.
121. Environmental Protection Act 1990. Section 6.
122. Environmental Protection Act 1990. Section 28(3)(a) and 28(4).
123. Council Directive 86/280/EEC of 12 June 1986 on limit values and quality objectives for discharges of certain dangerous substances included in List I of the Annex to Directive 76/464.
124. Unpublished personal correspondence.
125. National Rivers Authority v North West Water Plc. Darwen Magistrates Court Lancashire, 29 January 1992. Unreported.
126. Nigel Haigh. Manual of Environmental Policy p. 4.8–18. Longman.
127. Nigel Haigh. Manual of Environmental Policy p. 4.8–17. Longman.
128. Article 2(g) Council Directive 84/491/EEC of 9 October 1984 on limit values and quality objectives for discharges of hexachlorocyclohexane.
129. Nigel Haigh. Manual of Environmental Policy p. 4.8–18. Longman.

INDEX

For Product Safety Concerns and Information please contact our EU representative GPSR@taylorandfrancis.com Taylor & Francis Verlag GmbH, Kaufingerstraße 24, 80331 München, Germany

Batch number: 10732714

Printed by Printforce, the Netherlands